# ROAD MAINTENANCE AND REGRAVELLING
# USING LABOUR-BASED METHODS

## WORKBOOK

# Road maintenance and regravelling (ROMAR) using labour-based methods

# WORKBOOK

*Prepared for the International Labour Office*
*by*
CLAES-AXEL ANDERSSON
ANDREAS BEUSCH
and
DEREK MILES

INTERMEDIATE TECHNOLOGY PUBLICATIONS 1996

Intermediate Technology Publications Ltd
103–105 Southampton Row, London WC1B 4HH, UK

Copyright © International Labour Organization 1996

A CIP record of this book is available from
the British Library

ISBN 1 85339 349 5

Typeset by Dorwyn Ltd, Rowlands Castle, Hants
Printed by SRP, Exeter, UK

# CONTENTS

**SECTION B: PRACTICE**

# PREFACE

ROMAR stands for labour-based *Ro*utine *Ma*intenance and *R*egravelling, and extends the *Improve Your Construction Business (IYCB)* approach to small contractor development into this technical area. Readers who are familiar with the three IYCB handbooks and workbooks[1] will feel at home with the ROMAR handbook and workbook. As with IYCB, the ROMAR handbook provides ideas and information and the workbook gives readers a chance to look at their business in a disciplined way, and decide on action plans to make it more competitive and successful.

As more countries appreciate the social and economic benefits that can come from applying labour-based construction techniques, together with the productivity gains that can be achieved by entrusting this work to efficient private contractors, the market for ROMAR activities can be expected to grow. The ROMAR books will be particularly useful for experienced general contractors who are new to routine road maintenance and regravelling, and who need to judge whether their own resources and skills will enable them to turn this into an attractive business opportunity.

The ROMAR handbook and workbook are each divided into two parts, the first covering principles and the second dealing with the practice of labour-based road construction and maintenance. The 'route map' in the section on 'How to use your ROMAR books' will help you to find your way around, and concentrate on those chapters which are most relevant for your own circumstances. The books can be used in the context of an integrated training programme including opportunities to work on pilot contracts under supervision, but they have also been designed in a format suitable for self-study.

This book was prepared and edited within the Employment-Intensive Works Programme of the ILO based in its Development Policies Branch.

Claes-Axel Andersson<br>
Andreas Beusch<br>
Derek Miles

---

[1] Improve Your Construction Business series. No. 1: Pricing and bidding. No. 2: Site management. No. 3 Business management. ILO Geneva, 1994/6.

# THE AUTHORS

**Claes-Axel Andersson** is an independent consultant working with construction management issues, including small-scale contractor development. Until March 1996 he was an expert with the Employment-Intensive Works Programme of the ILO, based in its Development Policies Branch, also being responsible for ILO's Construction Management Programme including its Improve Your Construction Business activities. Mr. Andersson is a professionally qualified civil engineer with extensive experience in project management and building design.

**Andreas Beusch** is an independent consultant, specializing in appropriate road-work technology and training. He has over 15 years of experience, practical and managerial, of the development of labour-based construction methods in several developing countries. Mr. Beusch is a professionally qualified site engineer who started his career with his own planning and contracting firm.

**Derek Miles** is Director of the Institute of Development Engineering at Loughborough University, United Kingdom and is also Director of Overseas Activities in its Department of Civil and Building Engineering. He is a Fellow of the Institution of Civil Engineers and the Institute of Management and has more than 20 years' experience in the development of national construction industries. He directed the ILO Construction Management Programme during the period 1986–94.

# ACKNOWLEDGEMENTS

The Improve Your Construction Business (IYCB) approach to development of small-scale contractors was initiated through a pilot project in Ghana financed by the Government of the Netherlands. The IYCB concept has since been successfully introduced in other countries in Asia and Africa. The IYCB material, developed according to a modular concept, was already, at the conception stage, foreseen to be supplemented with training material covering technical and specific management issues for different construction sub-sectors.

In connection with a major World Bank infrastructure programme, the Government of Lesotho asked the ILO to undertake a contractor development project in the labour-based road sector. A major initial activity of this project was the production of a Road Maintenance and Regravelling (ROMAR) package expanding the IYCB concept into the labour-based road sector.

The ROMAR training sessions of the project were undertaken in close collaboration with the Labour Construction Unit (LCU) of the Lesotho Ministry of Works. The authors would like to recognize the dedication and enthusiasm of the LCU training staff that provided extensive assistance to the two rounds of contractor training. We would also like to recognize the two rounds of training provided comprehensive feed-back from trainees and trainers, which was incorporated when finalizing the manuscript.

# HOW TO USE YOUR ROMAR BOOKS

This book is written for you – the owner or manager of a small construction business. Together with the three basic IYCB handbooks and workbooks, the ROMAR books provide you with both management and technical advice that you need in order to make a success of routine maintenance and regravelling as a business activity. As with IYCB, the ROMAR handbook and workbook are best read together. We suggest you first read the chapter in the handbook, and then work through the examples in the corresponding chapter of the workbook.

## The handbook

The handbook, like the workbook, is divided into ten chapters. Part A, the first five chapters, deals with principles. The remainder of the book helps you to put these principles into practice.

Chapters 1 and 2 will be of most interest to those contractors who are new to the roads business. Chapter 1 summarizes the technical considerations, and also explains the most important engineering standards for earth and gravel roads. Chapter 2 provides an introduction to the choice of appropriate road construction and maintenance technology, and is written around a case study in which a new contractor looks at the business aspects of road works and decides on the type of equipment which will be required in order to carry out the various activities with a minimum capital outlay.

Chapter 3 introduces the reader to basic soil mechanics, including soil identification procedures, simple field tests and the principles of compaction. Chapter 4 deals with equipment, vehicles and tools, including repair and maintenance requirements, planning and reporting systems, service schedules and descriptions and basic specifications for common hand tools to assist in procurement. Chapter 5 completes the section on principles with a description of the major operations in a typical labour-based road project, and concludes with a case study drawn from one of the most successful labour-based programmes in Africa.

Once you are sure you have understood the principles set out in Part A, you will be ready to move on to Part B. Chapter 6 is a general chapter which explains how roads can deteriorate, the three main road maintenance systems, and the way in which maintenance contracts should be managed. If you are interested in routine maintenance you will then proceed to Chapter 7, which describes how to plan and carry out the 12 main routine maintenance activities. Chapter 8 has been written for those contractors who intend to specialize in regravelling, and describes how to plan and carry out these project-based operations. Even if you intend to specialize in only one of the two types of road maintenance, we recommend that you should look through both chapters before concentrating on the one in your own speciality area.

The book concludes with two chapters on ROMAR as a business activity, to supplement the topics dealt with in more general ways in the IYCB series. While the IYCB books relevant to labour-based road works go through pricing and bidding in detail, Chapter 9 in this book deals with special considerations relevant for road maintenance contracts, and shows how to build up detailed prices for a bill of quantities. Since labour-based projects inevitably mean that you have to cope with large workforces, Chapter 10 provides practical advice on how to manage people, including team building, communication, training, motivation, incentives, and how to ensure good discipline and morale. The book concludes with a list of reference literature and a summary of ILO Labour Standards.

# The workbook

The workbook enables you to test your understanding of ROMAR, and decide whether you have the knowledge, experience and resources to make a success of this business activity.

In each chapter of the workbook there is a list of simple questions to which you answer 'yes' or 'no'. The answers will tell you about the strengths and weaknesses of your existing business, and whether or not you should take the risk of venturing into one or both of the ROMAR activities (routine maintenance and/or regravelling).

If you have decided to go ahead, but still need to improve your knowledge in certain areas, you can then turn back to the appropriate section in the handbook and try again. Even when you have established yourself as a successful ROMAR contractor you will probably find it useful to go back to your handbook and workbook from time to time to see if there are ways to further improve your performance.

Please note that calculations are normally based on a number of assumptions, thus the answers to most calculations in both the handbook and the workbook are rounded to the nearest whole number or round figure, or to the nearest ten or hundred in the case of large numbers. This is done to avoid giving the impression that the end-result is more accurate than it actually is. Students using calculators should expect similar rather than identical answers in such estimates.

# Where to start

We recommend that you start by reading quickly through the whole of the handbook. Then you can go back over it more slowly, concentrating on the chapters which deal with topics which are new to you. As soon as you feel comfortable with the ideas presented in the handbook, you can try out your skills in the workbook.

---

Note: As this book is intended for use in many different countries, we have used the term 'NU' in the examples to represent an imaginary National Unit of currency, and 'NS' to stand for imaginary National Standards.

---

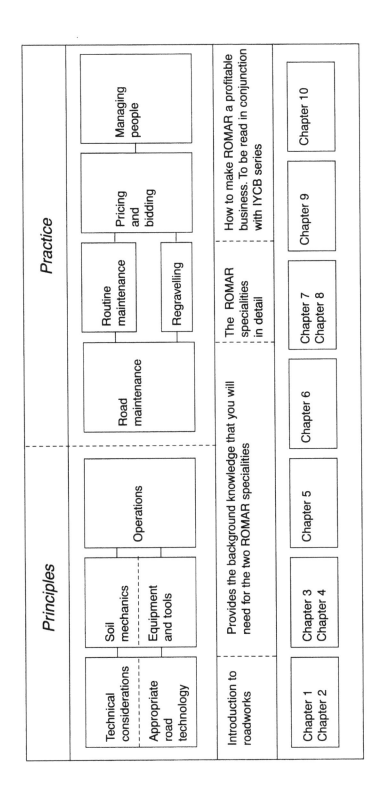

# SECTION A

# PRINCIPLES

# CHAPTER 1: ROADS: THEIR PURPOSE, TERMINOLOGY AND STANDARDS

## Quick Reference

Earth and Gravel Roads are the roads that you as a contractor will work on. In order to understand the contract documents fully and be able to communicate with the client it is important to know the common road terminology.

Technical standards are an important part of the contract documents. You must understand all the standards and instructions in the contract documents before you start working.

### REMEMBER

○ There are three principal types of roads: 1. paved roads, 2. gravel roads, 3. earth roads. Labour-based methods are best suited for gravel and earth roads
○ You must know all the road terms well enough to explain them to your employees in simple language.
○ Make sure that complete technical standards are included in the contract documentation.

## Part I – Business Questions

This section will help you to analyse how well you and your company are prepared to enter the labour-based road maintenance business. Go through the ten questions and answer all of them with yes or no. On the following page you can compare your answers with our checklist. In this checklist we help you to identify how you should prepare your company and yourself for this business opportunity.

|   | | Yes | No |
|---|---|:---:|:---:|

1. Do you know the approximate length of earth and gravel roads in your area of business operation? ☐ ☐

2. Do you have any practical experience of working on road construction or maintenance? ☐ ☐

3. Do you have any practical experience of constructing or maintaining drains or culverts? ☐ ☐

4. Do you know which authorities are responsible for road maintenance in your area, and whether they contract this work to the private sector? ☐ ☐

5. Have you studied the standard specifications for earth and gravel roads, and are you confident that you understand what resources you will need to keep to these standards for the full period of a routine maintenance contract? ☐ ☐

6. Do you understand that road maintenance will not bring high profits, but can cover basic overheads and yield steady earnings if it is well managed? ☐ ☐

7. Are you prepared to start with simple routine maintenance contracts, and then move on to regravelling and eventually new construction as your experience grows? ☐ ☐

8. Have you ever supervised more than ten manual labourers in your business activities? ☐ ☐

9. Do you feel confident that your workers will complete their allotted tasks without continuous supervision? ☐ ☐

10. Do you have experience in keeping records of work done, and submitting regular payment requests to your clients? ☐ ☐

## COMMENTS TO BUSINESS QUESTIONS

How many yes answers did you give? Multiply the number by ten, and check your percentage score. Your score will tell you how strong and well-prepared your company is. You may wish to look through the following checklist to help you understand why we think yes is the best answer to all these questions.

| | |
|---|---|
| 1. | You should know your potential market. |
| 2. and 3. | You are more likely to do well if you start with experience. |
| 4. | You need to know your client(s). |
| 5. | You must know what quality standards are expected. |
| 6. and 7. | You should not enter a new business activity with unrealistic expectations. |
| 8. and 9. | Managing people is a special skill. |
| 10. | You will lose money if you do not produce evidence to justify your requests for payment. |

# Part 2 – Business Practice

This section consists of two exercises that have been designed to test your understanding of road-work terminology and standards.

## EXERCISE 1: WHAT IS IT?

You are likely very soon to get a regravelling contract so you are preparing yourself and your company. One of the necessary activities is to brief your foreman on all important road terms, as he has never worked on roads before. Refer to the figures below and write the explanation you would give to your foreman on the lines below each picture.

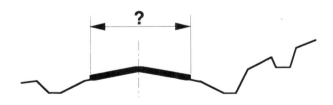

**Your Explanation to the Foreman**

. . . . . . . . . . . . . . . . . . . . . . . . . . . . . . . . . . . . . . . . . . . . . . . . . . . . . . . .
. . . . . . . . . . . . . . . . . . . . . . . . . . . . . . . . . . . . . . . . . . . . . . . . . . . . . . . .
. . . . . . . . . . . . . . . . . . . . . . . . . . . . . . . . . . . . . . . . . . . . . . . . . . . . . . . .
. . . . . . . . . . . . . . . . . . . . . . . . . . . . . . . . . . . . . . . . . . . . . . . . . . . . . . . .

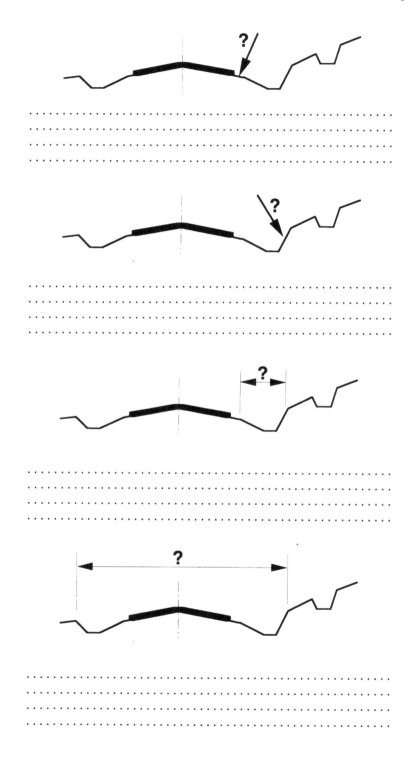

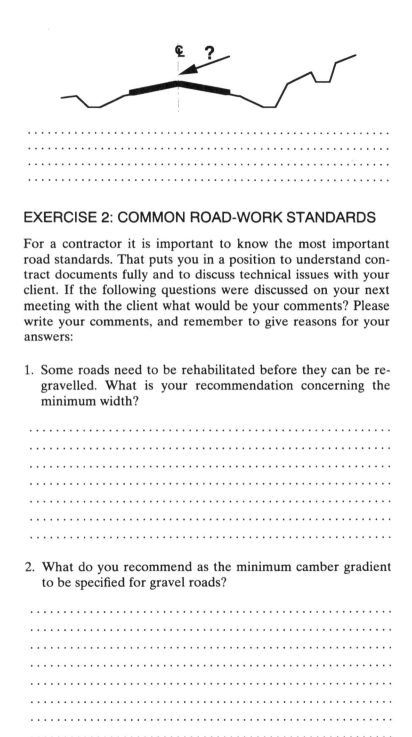

..................................................................
..................................................................
..................................................................
..................................................................

## EXERCISE 2: COMMON ROAD-WORK STANDARDS

For a contractor it is important to know the most important road standards. That puts you in a position to understand contract documents fully and to discuss technical issues with your client. If the following questions were discussed on your next meeting with the client what would be your comments? Please write your comments, and remember to give reasons for your answers:

1. Some roads need to be rehabilitated before they can be re-gravelled. What is your recommendation concerning the minimum width?

..............................................................
..............................................................
..............................................................
..............................................................
..............................................................
..............................................................
..............................................................

2. What do you recommend as the minimum camber gradient to be specified for gravel roads?

..............................................................
..............................................................
..............................................................
..............................................................
..............................................................
..............................................................
..............................................................
..............................................................

3. What is your recommendation concerning the shape and dimension of the side drains, bearing in mind that in future they will be maintained using labour-based methods (with lengthpersons).

. . . . . . . . . . . . . . . . . . . . . . . . . . . . . . . . . . . . . . . . . . . . . . . . . . . .
. . . . . . . . . . . . . . . . . . . . . . . . . . . . . . . . . . . . . . . . . . . . . . . . . . . .
. . . . . . . . . . . . . . . . . . . . . . . . . . . . . . . . . . . . . . . . . . . . . . . . . . . .
. . . . . . . . . . . . . . . . . . . . . . . . . . . . . . . . . . . . . . . . . . . . . . . . . . . .
. . . . . . . . . . . . . . . . . . . . . . . . . . . . . . . . . . . . . . . . . . . . . . . . . . . .
. . . . . . . . . . . . . . . . . . . . . . . . . . . . . . . . . . . . . . . . . . . . . . . . . . . .
. . . . . . . . . . . . . . . . . . . . . . . . . . . . . . . . . . . . . . . . . . . . . . . . . . . .
. . . . . . . . . . . . . . . . . . . . . . . . . . . . . . . . . . . . . . . . . . . . . . . . . . . .

## NOW CHECK YOUR ANSWERS

Our suggested answers are at the end of this chapter. We suggest that you check your answers against them before deciding on your action programme.

# Part 3 – Action Programme

## HOW TO CONSTRUCT YOUR ACTION PROGRAMME

Parts 1 and 2 should have helped you to understand your strengths and weaknesses as the owner or manager of a labour-based road maintenance enterprise. The general questions in Part 1 are a good guide to the strength of your business, and to the areas where there is most room for improvement. Look back at what percentage of 'yes' answers you had; the more yes answers, the more likely it is that your business will do well.

Now look again at those questions where you answered 'no'. These may be problem or opportunity areas for your business. Choose the one which you consider most important for your business at the present time. This is the sensible way to improve your business – take the most urgent problem first; don't try to solve everything at once.

Now write the problem or opportunity into the action programme chart, as we have done with the example. Then write in *What must be done, By whom* and *By when* in order to make sure things improve.

Finally, go back to your business and carry out the action programme.

| Problem | What must be done | By whom? | By when? |
|---|---|---|---|
| The only contract documentation I understand is that for building contracts. | Obtain standard conditions of contract from the roads authority, and compare this with documentation for building contracts. | Self/road authority | Next week |

# Answers to Business Practice

## EXERCISE 1: WHAT IS IT?

**Your Explanation to the Foreman**

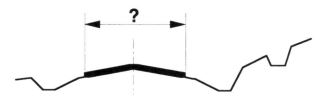

This is the width of the road which is paved, for example with gravel. It is called the **Carriageway**.

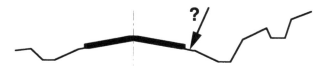

This is the **Road Shoulder** and it is the section of the road next to the carriageway. It ends where the ditch slope starts. The road shoulder can be paved or unpaved.

The **Back Slope** is the outer slope of the ditch. It is usually found on the hill-side of the road. The back slope should have an angle which prevents the soil from sliding into the ditch.

This is the **Side Drain**. It runs along the road side to collect the water flowing from the carriageway or nearby land. The side drain leads the water into mitre drians or culverts to dispose of it.

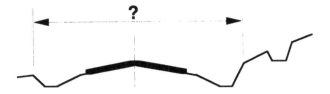

The full width of the road, including drains and fills is called the **Formation Width**.

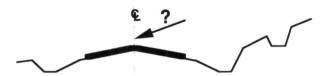

The **Road Centre Line** is the middle of the road. For setting out the road it is important to know where the centre line is.

## EXERCISE 2: COMMON ROAD-WORK STANDARDS

Our suggested answers:

1. We propose that the overall minimum width of the carriage-way, that is the width of the gravel layer, is not less than 3.50m. For safety reasons and for easy passage of vehicles this width is recommended as appropriate. The initial costs for the gravel layer will be slightly higher than for a width of 2.50m. However, in the long run costs for maintenance will be less as vehicles will not use the shoulder to pass each other.

2. The minimum camber gradient for gravel roads should not be less than 5% when compacted. However, we also recommend that the maximum gradient should not exceed 8% as this would lead to erosion of the gravel layer.

3. As the intention is to improve and maintain the roads using labour-based methods, it is recommended to specify that the ditches be of trapezoidal shape. The minimum width at the bottom should not be less than 40cm.

# CHAPTER 2: APPROPRIATE ROAD CONSTRUCTION AND MAINTENANCE TECHNOLOGY

## Quick reference

In many countries road-making equipment is expensive to buy and maintain, but the lack of alternative employment opportunities means that wage rates are low and casual workers can be recruited as needed. If your country fits this pattern, it makes sense to substitute labour for equipment whenever you can. The development of labour-based construction technologies over recent years offers contractors new opportunities to reduce costs and avoid the financial risk of buying expensive plant which is idle between contracts.

Labour-based regravelling is a good business for the contractor with limited capital, since you only need simple equipment like a pedestrian roller, a water bowser, a tractor and trailer for transport and a pick-up for supervision. Besides being cheap and reliable, a tractor and trailer can often be hired out to local farmers between contracts to provide another source of income. Routine maintenance also needs very little start-up capital, because the main costs are labour wages and hand tools. Another advantage of annual routine maintenance contracts is that they provide regular income, which compensates for the risks of project contracting.

### REMEMBER

○ Practical utilization rates for heavy construction plant are often as low as 30 per cent in developing countries, so theoretical plant productivity levels should not be used in plant purchase calculations unless you are really sure that they will apply in your own situation.
○ Labour-based techniques can be more profitable than equipment-based techniques for contractors with good management, but you need well-trained and well-motivated supervisors to get the best out of your labour force.

# Part 1 – Business Questions

This section will help you to analyse how well prepared you and your company are to enter the labour-based road maintenance business. Go through the ten questions and answer all of them with yes or no. Then you can compare your answers with our checklist. In this checklist we help you to identify how you should prepare your company and yourself for this business opportunity.

|  | Yes | No |
|---|---|---|
| 1. Can you list all the common operations that are required in a routine maintenance contract? | ☐ | ☐ |
| 2. Can you list all the common operations that are required in a regravelling contract? | ☐ | ☐ |
| 3. Do you regularly look at common operations required in your contracts, and compare the costs of carrying them out by labour-based and equipment-based methods? | ☐ | ☐ |
| 4. Do you understand the importance of checking on true plant utilization rates in calculating equipment costs? | ☐ | ☐ |
| 5. Do you use daily and weekly plant return forms to check regularly on the *true* utilization rates for your own equipment? | ☐ | ☐ |
| 6. Before buying new or second-hand equipment, do you always try to calculate its likely working life and the cost of keeping it in good repair? | ☐ | ☐ |
| 7. Do you prefer to buy general-purpose equipment (such as tractors and trailers), rather than specialist heavy equipment that can only be used for a few tasks on occasional contracts? | ☐ | ☐ |
| 8. Do you take care to recruit supervisors who have the ability to achieve high quality work and meet productivity targets? | ☐ | ☐ |

|  | Yes | No |
|---|:---:|:---:|

9. Do you give clear instructions to your supervisors, and check regularly on performance?

10. Do you treat your supervisors as key staff, and provide them with training and financial incentives?

## COMMENTS TO BUSINESS QUESTIONS

How many yes answers did you give? Multiply the number by ten, and check your percentage score. Your score will tell you how well prepared your company is for this business opportunity. You may wish to look through the following checklist to help you understand why we think yes is the best answer to all these questions.

| | |
|---|---|
| 1. and 2. | You must be able to list all the operations so as to plan and cost the work. |
| 3. | Choosing the right technology can save you money. |
| 4. and 5. | Do not rely upon productivity figures or costs given in textbooks or manufacturers' catalogues without checking whether they apply in your own situation. |
| 6. and 7. | It is only worth buying or hiring plant if you are sure that you can keep it working. |
| 8., 9. and 10. | Labour-based construction activities can be profitable providing they are properly supervised, so you must be sure to recruit trustworthy supervisors and look after them well. |

# Part 2 – Business Practice

This section consists of two exercises that have been designed to help you choose the right technology approach for road-works as a small-scale contractor.

## EXERCISE 1: GRADER OR LABOUR?

You have just won your first road contract; the regravelling of 10km of road in a rural area. The client has placed no restrictions on the type of technology you can use, so you have the freedom to make your own decisions. So far you have decided to use labour to excavate, load and spread the gravel, but haulage will be done using your own two tipper trucks plus another two which can be hired locally.

You still have to decide how to reshape the road to the design standard, and the choice is between labour-based methods and a grader. If you already owned a grader you would probably use it, but you have to decide whether it is worth buying or hiring one. You could not afford to buy a new grader at a cost of about 140 000 NU, but a second-hand one is available for about 80 000 NU from an international contractor who has just finished a large project. You calculate that the component costs of depreciation, interest and running costs* would result in an hourly cost of 35 NU using a realistic annual utilization rate of 30 per cent.

The alternative is to hire a grader from a local dealer at an inclusive hourly rate of 50 NU. You estimate that the grader could complete some 2.5km of reshaping work in a day, including reshaping the camber and opening the ditches.

As this is an area with high unemployment, you also have access to as many labourers as you need at a daily wage rate of 1.70 NU. From experience you know that a labourer can reshape 20 metres of road per day and achieve the quality of work specified in the contract documents.

How do you propose to carry out the reshaping work?

## EXERCISE 2: WHAT DO I NEED?

You and your partner are the owners of a successful building business, and you own some basic plant including a flat-bed truck, two concrete mixers, a vibrator, a dumper, some scaffolding and a pick-up. Your contracts have been profitable and you have just over 100 000 NU in your joint bank account. You feel the time has come to diversify your business interests, and are considering branching out into labour-based regravelling.

You have contacted the local roads authority, and have been told that you will be considered for contracts providing you own the basic equipment necessary to carry out regravelling contracts of 10–15km. You obtain the following equipment list from the local plant dealer. What do you propose to invest in at this stage?

---

* See IYCB Handbook 3 for details of how to calculate plant costs.

| Item | Unit price (NU) | Number required | Total (NU) |
|---|---|---|---|
| 7 ton tipper | 60 000 | | |
| Dozer (D6) | 150 000 | | |
| Grader (120G) | 138 000 | | |
| Front-end loader | 146 000 | | |
| Tractor (60hp) | 25 000 | | |
| Trailer (3.5 cu.m.) | 6000 | | |
| Vibrating roller, self-propelled | 65 000 | | |
| Pedestrian vibrating roller | 5000 | | |
| Pick-up (1.5 ton) | 35 000 | | |
| Saloon car (BMW) | 28 000 | | |
| Flat-bed truck | 50 000 | | |
| Total | | | |

## NOW CHECK YOUR ANSWERS

Our suggested answers are at the end of this chapter. We suggest you check your answers against them before deciding on your action programme.

# Part 3 – Action Programme

## HOW TO CONSTRUCT YOUR ACTION PROGRAMME

Parts 1 and 2 should have helped you to understand your strengths and weaknesses as the owner or manager of a labour-based road maintenance enterprise. The general questions in Part 1 are a good guide to the strength of your business, and to the areas where there is most room for improvement. Look back at what percentage of 'yes' answers you had; the more yes answers, the more likely it is that your business will do well.

Now look again at those questions where you answered 'no'. These may be problem or opportunity areas for your business. Choose the one which you consider most important for your business at the present time. This is the sensible way to improve your business – take the most urgent problem first; don't try to solve everything at once.

Now write the problem or opportunity into the action programme chart, as we have done with the example. Then write *What must be done, By whom* and *By when* in order to make things improve.

Finally, go back to your business and carry out the action programme.

| Problem | What must be done | By whom? | By when? |
|---------|-------------------|----------|----------|
| I have no supervisors with labour-based roads experience. | Either send one of my building supervisors on a course run by the roads authority or recruit a retired roads supervisor with labour-based experience. First step is to ask for advice from the Regional Engineer. | Me | End of this month |

# Answers to Business Practice

### EXERCISE 1: GRADER OR LABOUR?

There are three possible ways of reshaping the road:
buy a second-hand grader (calculated hourly cost 35 NU)
hire a grader at a rate of 50 NU per hour
Use labour-based methods

*Option 1: Buy a grader*
The daily output is estimated to be 2500m and the daily operational costs are 35 NU multiplied by 8 hours = 280 NU. To carry out the whole reshaping operation the grader would take only four working days and would cost 1120 NU.

*Option 2: Hire a grader*
The daily output is again 2500m, so the daily operational costs are 50 NU multiplied by 8 hours = 400 NU. To carry out the whole reshaping operation the grader would take only 4 working days and would cost 1600 NU.

*Option 3: Labour-based*
The daily output per labourer is 20m at a cost of 1.70 NU. The total cost for labour wages would be 1.70 NU, divided by 20m, multiplied by 10 000m = 850 NU. An additional 8 per cent is added to cover the cost of hand tools (68 NU) and 20 per cent for overheads (184 NU). Thus the total costs of labour would be approximately 1100 NU.

*Your choice*
Clearly it would not pay to hire the grader, as the cost would be half as much again as that for labour-based methods. This means that the choice lies between labour-based methods and buying the grader. The calculated costs are approximately equal, so there is little to choose between them on mathematical grounds.

If you had plenty of money and you were sure of a steady workload, the chance to buy a second-hand grader might be tempting. But you should also take account of risk. If subsequent contracts are not awarded, you will own an expensive piece of machinery that is not suitable for other work. If the second-hand grader breaks down your work will be held up and the cost of repairs may be higher than you have estimated. Contracting is always a risky activity, so you should probably play safe and use labour-based methods and make a slightly better profit without making an expensive and risky investment.

## EXERCISE 2: WHAT DO I NEED?

You have decided to concentrate on labour-based regravelling and maintenance, rather than new construction. You already have a pick-up, so there is no point in buying another until you really need it. There is also no point in spending all your company's hard-earned savings of 100 000 NU on equipment that is not necessary at this stage. So you will probably choose only what is essential in order to convince the client that you are serious.

The only equipment you actually need is for the hauling of gravel, compaction of the gravel layer and for supervision of the regravelling works as well as for routine maintenance supervision. Your choice will depend on local conditions, but we have suggested that you might buy two tractors and four trailers (to give enough capacity to achieve reasonable outputs on transporting material), plus two pedestrian vibrating rollers (which will provide matching capacity for compaction, and will provide some back-up if one of the rollers needs repair).

| Item | Unit price (NU) | Number required | Total (NU) |
|------|-----------------|-----------------|------------|
| 7 ton tipper | 60 000 | | |
| Dozer (DA) | 150 000 | | |
| Grader (120G) | 138 000 | | |
| Front-end loader | 146 000 | | |
| Tractor (60hp) | 25 000 | 2 | 50 000 |
| Trailer (3.5 cu.m.) | 6000 | 4 | 24 000 |
| Vibrating roller, self-propelled | 65 000 | | |
| Pedestrian vibrating roller | 5000 | 2 | 10 000 |
| Pick-up (1.5 ton) | 35 000 | | |
| Saloon car (BMW) | 28 000 | | |
| Flat-bed truck | 50 000 | | |
| TOTAL | | | 84 000 |

19

# CHAPTER 3:   SOIL MECHANICS

## Quick Reference

There are two main reasons for compacting earth or gravel to form the surface of a road:

○ to spread the traffic load; and
○ to prevent water from penetrating the road layers and weakening them.

On lightly trafficked roads compacted natural soil is often sufficient, but road gravel (which is more cohesive than builder's gravel used in concreting) is imported from borrow pits to provide a stronger surface on more heavily trafficked roads. The process of transporting, placing and compacting gravel is known as 'gravelling' or 'regravelling'.

Earth and gravel are the key materials for the road contractor, so you need to know how to recognize them, how to test them and how to compact them. Soil is a mixture of solid particles, water and air. Compaction is a process which pushes the particles closer together by squeezing out the air so as to make the material stronger and more stable. It is usually done on small contracts by towed dead-weight rollers or pedestrian vibrating rollers.

Simple field tests (mainly bottle tests) are usually sufficient to check whether material can be used for gravelling work or not. If these tests show that the identified gravel is of doubtful quality, you should inform the client immediately.

### REMEMBER

○ Gravel for road surfaces has to be well graded. This means that the stone content should be about 50 per cent, the sand content about 40 per cent and the clay content about 10 per cent.
○ Settling (or bottle) tests allow the identification of the approximate content of the different soil fractions. Moulding tests or drying tests can be used to determine the proportion of clay and silt.
○ Laboratory tests will always be necessary when detailed soil classification is required.
○ Compaction achieves the best results if the gravel has the optimum moisture content (8 to 10 per cent of total volume).

# Part 1 – Business Questions

This section will help you to analyse how well prepared you and your company are to enter the labour-based road maintenance business. Go through the ten questions and answer all of them with yes or no. Then you can compare your answers with our checklist. In this checklist we help you to identify how you should prepare your company and yourself for this business opportunity.

|  | Yes | No |
|---|---|---|
| 1. When you visit a construction site, can you tell whether the soil is gravel, sand, silt, clay or organic? | ☐ | ☐ |
| 2. Do you know how to test whether a soil sample is well graded, poorly graded or uniformly graded? | ☐ | ☐ |
| 3. Can you draw a diagram which explains the function of a gravel road surface? | ☐ | ☐ |
| 4. Can you draw a diagram which illustrates the USCS soil identification procedure? | ☐ | ☐ |
| 5. Have you ever carried out a field settling (bottle) test? | ☐ | ☐ |
| 6. Have you ever carried out a moulding test? | ☐ | ☐ |
| 7. Have you ever carried out a drying (matchbox) test? | ☐ | ☐ |
| 8. Could you explain to your site supervisor the reason why it is important to achieve good compaction? | ☐ | ☐ |
| 9. Could you explain to your site supervisor the reason why it is important to compact soil close to the optimum moisture content? | ☐ | ☐ |
| 10. Have you had field experience of all the main ways of compacting gravel layers, and do you know how to compare the costs of achieving a given bearing capacity? | ☐ | ☐ |

## COMMENTS TO BUSINESS QUESTIONS

How many yes answers did you give? Multiply the number by ten, and check your percentage score. You may look through the following checklist to help you understand why we think yes is the best answer to all these questions.

| | |
|---|---|
| 1. and 2. | For a road contractor, soil is an important material. You need to understand what types of soil are suitable for construction purposes. |
| 2., 3. and 4. | Although you do not need to be an expert in soil mechanics, you need to understand the basic principles so you can explain them to your supervisors. |
| 5., 6. and 7. | As a contractor, you should be able to carry out basic tests to make sure that your materials meet the quality standards specified by the client. |
| 8., 9. and 10. | Compaction is a vital technical process in road construction and maintenance. If it is not done properly you will lose both your profit and your reputation. |

# Part 2 – Business Practice

This section consists of two exercises designed to test your knowledge of different soils and their suitability for road construction purposes.

## EXERCISE 1: CAN THIS SOIL BE USED AS SURFACE MATERIAL?

Your company has won a gravelling contract for which the quarry has been located by the client. The contract specifies the gravel to be used as 'GC, well graded' in accordance with the USCS classification system. At the design stage the client dug sample pits in the quarry and sent the material to a laboratory for testing. The results were according to the required specifications.

You have now removed all the topsoil from the quarry and you are doubtful whether the gravel from all quarry locations is as good as the tested samples. In one corner the gravel seems to be distinctly different. You have therefore decided to carry out your own field tests, because you know that you will finally be responsible for the quality of the surface material.

After the soil fractions have settled in the jar, they look like the figure opposite.

How would you classify this sample? Do you think that it is safe to use this material for regravelling?

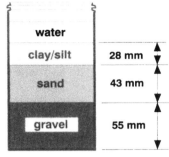

## EXERCISE 2:
## TERMINOLOGY FOR SOIL MECHANICS

It is important for you to understand the meaning of the most common terms in soil mechanics. In this exercise you have to choose the best description for various types of soil, and terms used to describe soils:

*1. Organic soil*
a. Ideal material for regravelling.
b. Topsoil.
c. A coarse to fine gritty soil.
d. Soil with small particles, powdery when dry but soft when wet.
e. Soil with very small particles, lumpy when dry but sticky and soft when wet.

*2. Well-graded material*
a. Ideal material for regravelling.
b. Topsoil.
c. A coarse to fine gritty soil.
d. Soil with small particles, powdery when dry but soft when wet.
e. Soil with very small particles, lumpy when dry but sticky and soft when wet.

*3. Clay*
a. Ideal material for regravelling.
b. Topsoil.
c. A coarse to fine gritty soil.
d. Soil with small particles, powdery when dry but soft when wet.
e. Soil with very small particles, lumpy when dry but sticky and soft when wet.

*4. Silt*
a. Ideal material for regravelling.
b. Topsoil.
c. A coarse to fine gritty soil.
d. Soil with small particles, powdery when dry but soft when wet.
e. Soil with very small particles, lumpy when dry but sticky and soft when wet.

5. *Sand*
a. Ideal material for regravelling.
b. Topsoil.
c. A coarse to fine gritty soil.
d. Soil with small particles, powdery when dry but soft when wet.
e. Soil with very small particles, lumpy when dry but sticky and soft when wet.

6. *Compaction*
a. The strength of the soil.
b. Measures whether soil can be moulded and hold its new shape.
c. A process which packs particles closer together.
d. The degree to which water can penetrate a soil.
e. The water `content that gives the best effect of soil compaction.

7. *Optimum moisture content*
a. The strength of the soil.
b. Measures whether soil can be moulded and hold its new shape.
c. A process which packs particles closer together.
d. The degree to which water can penetrate a soil.
e. The water content that gives the best effect of soil compaction.

8. *Permeability*
a. The strength of the soil.
b. Measures whether soil can be moulded and hold its new shape.
c. A process which packs particles closer together.
d. The degree to which water can penetrate a soil.
e. The water content that gives the best effect of soil compaction.

9. *Bearing capacity*
a. The strength of the soil.
b. Measures whether soil can be moulded and hold its new shape.
c. A process which packs particles closer together.
d. The degree to which water can penetrate a soil.
e. The water content that gives the best effect of soil compaction.

*10. Plasticity*
a. The strength of the soil.
b. Measures whether soil can be moulded and hold its new shape.
c. A process which packs particles closer together.
d. The degree to which water can penetrate a soil.
e. The water content that gives the best effect of soil compaction.

## NOW CHECK YOUR ANSWERS

Our suggested answers at the end of this chapter. We suggest you check your answers against them before deciding on your action programme.

# Part 3 – Action Programme

## HOW TO CONSTRUCT YOUR ACTION PROGRAMME

Parts 1 and 2 should have helped you to understand your strengths and weaknesses as the owner or manager of a labour-based road maintenance enterprise. The general questions in Part 1 are a good guide to the strength of your business, and to the areas where there is most room for improvement. Look back at what percentage of 'yes' answers you had; the more yes answers, the more likely it is that your business will do well.

Now look again at those questions where you answered 'no'. These may be problem or opportunity areas for your business. Choose the one which you consider most important for your business at the present time. This is the sensible way to improve your business – take the most urgent problem first; don't try to solve everything at once.

Now write the problem or opportunity into the action programme chart, as we have done with the example. Then write *What must be done, By whom* and *By when* in order to make sure things improve.

Finally, go back to your business and carry out the action programme.

| Problem | What must be done | By whom? | By when? |
|---|---|---|---|
| I do not think my supervisors are carrying out field tests properly. | I have never carried out field tests myself, so I cannot be sure that my suspicions are correct. I will take advice from a local engineer, and practice carrying out tests at home. | Myself plus engineer. | This weekend. |

# Answers to Business Practice

EXERCISE 1: CAN THIS SOIL BE USED AS SURFACE MATERIAL?

The first step is to calculate the percentages of the vairous fractions. Since the cross-section of the bottle is uniform, the proportions by volume will be based on their heights as measured with a ruler (indicated on Figure 3.1):

| | |
|---|---|
| Clay/silt | 28mm |
| Sand | 43mm |
| Gravel | 55mm |
| Total | 126mm |

The percentages are calculated by dividing the heights of the three fractions by 126, and then multiplying by 100:

| | | |
|---|---|---|
| Clay silt | $28/126 = 0.22, 100 \times 0.22 =$ | 22 |
| Sand | $43/126 = 0.34, 100 \times 0.34 =$ | 34 |
| Gravel | $55/126 = 0.44, 100 \times 0.44 =$ | 44 |
| Total | | 100 |

The soil identification is now carried out in three stages:

1. The percentage of the gravel sand layer is 78 per cent, so according to the classification table this is a **coarse grained soil**.
2. We now need to know whether the sample is gravel or sand. Therefore we calculate the fractions of gravel and sand in relation to the total height. The result is gravel 44 per cent, sand 34 per cent. Although the gravel content is not over 50 per cent, this sample must be classified as **gravel** since there is clearly more gravel than sand in the sample.
3. After identifying it as gravel we want to know whether the sample has sufficient or too much fines (clay and/or silt). The ideal mixture should have about 10 to 12 per cent clay in relation to the total sample height amounts to 22 per cent. According to the soil classification table (Handbook Table 3.1) we now know that we have **dirty gravel**. However, the sample is not well-graded as the content of fines is definitely above the tolerated limit.

*Decision*

It is not necessary to carry out the next test to establish whether the fines consist of silt or clay as the fines content is too high anyway. As a result of your tests you should inform the client in writing about your findings and recommend that further samples of the gravel from the particular corner of the quarry be tested in a laboratory to identify the precise quality of the gravel.

As the quarry was specified by the client, you should also ask the client to inform you in writing whether the gravel should be used or not, as soon as the laboratory tests have been carried out. If the client still wants to use the gravel, you will then not be liable for a reduced bearing-capacity or increased loss of gravel at a later stage.

Finally, you should consider whether you are entitled to make a claim, under the terms of the contract, to cover the costs resulting from delays while the matter is resolved.

## EXERCISE 2: TERMINOLOGY FOR SOIL MECHANICS

Our suggested answers are as follows:

1. b.
2. a.
3. e.
4. d.
5. c.
6. c.
7. e.
8. d.
9. a.
10. b.

# CHAPTER 4   EQUIPMENT AND TOOLS

## Quick Reference

Equipment and vehicles for labour-based road works are usually only used for activities where labour would significantly delay completion or where the quality requirements demand equipment. To make the best use of your equipment and reduce the number of breakdowns, appropriate maintenance and repair arrangements have to be made. You can either set up a service unit within your own company, or have a service contract with an outside agency or garage.

As a labour-based contractor you will need to keep a large stock of hand tools. It pays to think carefully about which tools to buy. Normally it is better in the long run to buy tools of good quality, even if they are more expensive.

In order to keep track of your equipment and tools, it is essential to set up a simple equipment planning and reporting system. You should also use the information to monitor and analyse operating costs, average availability and utilization, fuel consumption, spare part consumption and condition of the equipment. This will help you to find out where you can cut costs, as well as improving decisions about new purchases.

### REMEMBER

○ It often pays to have a stock of 'fast-moving spare parts' at site to ensure good equipment utilization.
○ Your labour is more productive if provided with appropriate hand tools of adequate quality.
○ Agricultural hand tools are often not of sufficiently high quality for road construction.

# Part 1 – Business Questions

This section will help you to analyse how well prepared you and your company are to enter the labour-based road maintenance business. Go through the ten questions and answer all of them with yes or no. Then you can compare your answers with our checklist. In this checklist we help you to identify how you should prepare your company and yourself for this business opportunity.

|  | Yes | No |
|---|---|---|
| 1. Have you calculated the minimum investment that you will have to make in basic equipment and vehicles in order to operate as a regravelling contractor? | ☐ | ☐ |
| 2. Do all your workers have the hand tools that they need in order to carry out their work? | ☐ | ☐ |
| 3. Do you carry out regular site inspections of hand tools, to make sure that they are kept clean and in good condition? | ☐ | ☐ |
| 4. If you rely on hiring for part of your equipment needs, do you always check on local sources of supply (and hire rates!) before bidding? | ☐ | ☐ |
| 5. Do you try to standardize on a single make of basic equipment, like tractors and rollers? | ☐ | ☐ |
| 6. Do you make sure that you mechanics are properly trained by the supplier of the equipment? | ☐ | ☐ |
| 7. Do you keep a stock of basic spares on remote sites? | ☐ | ☐ |
| 8. Do you keep accurate cost records for all large items of plant? | ☐ | ☐ |
| 9. Do you make sure that drivers and plant operators undertake basic daily maintenance? | ☐ | ☐ |
| 10. Do you make sure that your mechanics keep maintenance records for all major plant, to ensure that routine maintenance takes place as recommended by the manufacturer? | ☐ | ☐ |

## COMMENTS TO BUSINESS QUESTIONS

How many yes answers did you give? Multiply the number by ten, and check your percentage score. Your score will tell you how strong and well prepared your company is. You may wish to look through the following checklist to help you understand why we think yes is the best answer to all these questions.

| | |
|---|---|
| 1. | You should not enter a new business activity unless you are sure that you have enough fixed capital *and* enough working capital (see IYCB Handbook 3). |
| 2. and 3. | You will only get good productivity from your workers if they have the right tools and they keep them in good condition. |
| 4. | Hire rates vary from place to place, and you could lose a lot of money if local rates turn out to be higher than expected. |
| 5., 6. and 7. | Try to avoid equipment downtime by standardization employing trained mechanics and keeping a stock of frequently-used replacement items. |
| 8. | Equipment is expensive to buy and repair, so effective cost control is essential. |
| 9. and 10. | Regular daily and periodic routine maintenance reduces downtime and minimizes costs. |

# Part 2 – Business Practice

## WHICH TOOLS ARE NEEDED?

When planning site activities, it is essential to know what tools will be required to execute the work. In employment-intensive road works, where almost all activities are carried out by labour, it is of course even more crucial. The tools that are commonly used in labour-based road works are listed below:

*List of tools employed on labour-based sites*

| | |
|---|---|
| ○ Pickaxe | ○ Spreader |
| ○ Axe | ○ Sledge-hammer |
| ○ Mattock | ○ Crowbar |
| ○ Shovel | ○ Earth rammer |
| ○ Spade | ○ Bow saw |
| ○ Long-handled spade | ○ Bushman saw |
| ○ Long-handled spike | ○ Grass slasher |
| ○ Hoe | ○ Bush knife |
| ○ Forked Hoe | ○ Grass cutter |
| ○ Rake | ○ Wheelbarrow |

For the exercises in this chapter, we assume a routine maintenance contract is out for tender. We look at each of the different activities we will be doing. Before you can identify the tools required for each activity a site inspection has to be undertaken. The information provided under each question are the notes you made during the site visit. You should choose from the above list the appropriate tools for each activity.

## EXERCISE 1

*Activity:* Inspection and removal of obstructions and debris over the entire road length.

*Observations:* The road runs partially through thick forest where occasionally a tree or branches could fall on to the road.

*Tools required:* ........................................
................................................

## EXERCISE 2

*Activity:* Clear silt and debris from culvert, inlet and outlet ditch, and dispose the material behind the side drain.

*Observations:* Culvert length is approx. 7.5m, some culvert outlets are overgrown with vegetation.

*Tools required:* ........................................
................................................

## EXERCISE 3

*Activity:* Clean the side drains to the standard cross-section removing all soil, vegetation and other debris and dispose the material inside or outside the road reserve.

*Observations:* Most drains are silted and the silt seems to be rather soft.

*Tools required:* ........................................
................................................

## EXERCISE 4

*Activity:* Repair scour checks to standard dimensions using stones or wooden sticks.

*Observations:* Stones are not available along this road but there is plenty of wood.

*Tools required:* . . . . . . . . . . . . . . . . . . . . . . . . . . . . . . . . . . . . . . . . . . . .
. . . . . . . . . . . . . . . . . . . . . . . . . . . . . . . . . . . . . . . . . . .

## EXERCISE 5

*Activity:* Clean the mitre drains to the standard cross-section removing all soil, vegetation and other debris and dispose the material inside or outside the road reserve.

*Observations:* Some of the mitre drains are overgrown with vegetation.

*Tools required:* . . . . . . . . . . . . . . . . . . . . . . . . . . . . . . . . . . . . . . . . . . . .
. . . . . . . . . . . . . . . . . . . . . . . . . . . . . . . . . . . . . . . . . . .

## EXERCISE 6

*Activity:* Fill pot-holes with approved material and compact well.

*Observations:* The gravel layer seems to be very hard. Gravel for pot-hole filling has been provided and stacked along the road. The gravel is relatively loose.

*Tools required:* . . . . . . . . . . . . . . . . . . . . . . . . . . . . . . . . . . . . . . . . . . . .
. . . . . . . . . . . . . . . . . . . . . . . . . . . . . . . . . . . . . . . . . . .

## EXERCISE 7

*Activity:* Grub carriageway edge and remove roots and spoil well clear of the carriageway on to the road reserve.

*Observations:* The roots of the grass do not appear to be very deep or difficult to grub.

*Tools required:* . . . . . . . . . . . . . . . . . . . . . . . . . . . . . . . . . . . . . . . . . . . .
. . . . . . . . . . . . . . . . . . . . . . . . . . . . . . . . . . . . . . . . . . .

## EXERCISE 8

*Activity:* Repair shoulder and slope erosion.

*Observations:* Only few erosion gullies. Grass to replant is available in the road reserve.

*Tools required:* ........................................
........................................

## EXERCISE 9

*Activity:* Cut grass to a maximum height of 50mm above ground level on shoulders, ditch bottoms and side slopes.

*Observations:* No special problems.

*Tools required:* ........................................
........................................

## EXERCISE 10

*Activity:* Clear bush and remove roots and stumps from side ditch, shoulder, slope and road reserve 2m from outer side of ditch.

*Observations:* On some sections the road reserve is overgrown with heavy bush and even small trees.

*Tools required:* ........................................
........................................

## NOW CHECK YOUR ANSWERS

Our suggested answers are at the end of this chapter. We suggest you check your answers against them before deciding on your action programme.

# Part 3 – Action Programme

## HOW TO CONSTRUCT YOUR ACTION PROGRAMME

Parts 1 and 2 should have helped you to understand your strengths and weaknesses as the owner or manager of a labour-based road maintenance enterprise. The general questions in Part 1 are a good guide to the strength of your business, and to the areas where there is most room for improvement. Look back at what percentage of 'yes' answers you had; the more yes answers, the more likely it is that your business will do well.

Now look again at those questions where you answered 'no'. These may be problem or opportunity areas for your business. Choose the one which you consider most important for your business at the present time. This is the sensible way to improve your business – take the most urgent problem first; don't try to solve everything at once.

Now write the problem or opportunity into the action programme chart, as we have done with the example. Then write in *What must be done, By whom* and *By when* in order to make sure things improve.

Finally, go back to your business and carry out the action programme.

| Problem | What must be done | By whom? | By when? |
|---|---|---|---|
| I am always buying tools for my sites. It is costing me a fortune. | Set up a tool registration system at each site, and make sure that the site supervisor makes a regular inventory. | Me, and site supervisor | System in operation from next month |

# Answers to Business Practice

## EXERCISE 1

*Activity:* Inspection and removal of obstructions and debris over the entire road length.

*Observations:* the road runs partially through thick forest where occasionally a tree or branches could fall on to the road.

*Tools required:* Rake, Axe, Bushman saw

## EXERCISE 2

*Activity:* Clear silt and debris from culvert, inlet and outlet ditch, and dispose the material behind the side drain.

*Observations:* culvert length is approx. 7.5m, some culvert outlets are overgrown with vegetation.

*Tools required:* Long-handled spade, long-handed spike, Hoe, Shovel

## EXERCISE 3

*Activity:* Clean the side drains to the standard cross-section removing all soil, vegetation and other debris and dispose the material inside or outside the road reserve.

*Observations:* most drains are silted and the silt seems to be rather soft.

*Tools required:* Hoe, Shovel

## EXERCISE 4

*Activity:* Repair scour checks to standard dimensions using stones or wooden sticks.

*Observations:* stones are not available along this road but there is plenty of wood.

*Tools required:* Axe, Bush knife, Pick axe, Hoe, Shovel, Sledgehammer, Wheelbarrow

## EXERCISE 5:

*Activity:* Clean the mitre drains to the standard cross-section removing all soil, vegetation and other debris and dispose the material inside or outside the road reserve.

*Observations:* some of the mitre drains are overgrown with vegetation.

*Tools required:* Grass slasher, Bush knife, Hoe, Shovel

## EXERCISE 6

*Activity:* Fill pot-holes with approved material and compact well.

*Observations:* the gravel layer seems to be very hard. Gravel for pot-hole filling has been provided and stacked along the road. The gravel is relatively loose.

*Tools required:* Wheelbarrow, Hoe, Shovel, Rake, Spreader, Earth rammer

## EXERCISE 7

*Activity:* Grub carriageway edge and remove roots and spoil well clear of the carriageway on to the road reserve.

*Observations:* the roots of the grass do not appear to be very deep or difficult to grub.

*Tools required:* Hoe, Shovel, Rake

## EXERCISE 8

*Activity:* Repair shoulder and slope erosion.

*Observations:* only few erosion gullies. Gradd to re-plant is available in the road reserve.

*Tools required:* Hoe, Shovel, Rake, Earth rammer, Wheelbarrow

## EXERCISE 9

*Activity:* Cut grass to a maximum height of 50mm above ground level on shoulders, ditch bottoms and side slopes.

*Observations:* no special problems.

*Tools required:* Grass slasher

## EXERCISE 10

*Activity:* Clear bush and remove roots and stumps from side ditch, shoulder, slope and road reserve 2m from outer side of ditch.

*Observations:* On some sections the road reserve is overgrown with heavy bush and even small trees.

*Tools required:* Bushman saw, Axe, Bush knife, Hoe, Mattock, Shovel, Wheelbarrow

# CHAPTER 5: INTRODUCTION TO LABOUR-BASED ROAD CONSTRUCTION

## Quick Reference

To implement labour-based road works effectively, with up to 200 labourers on a project, you need an efficient work organization. Good planning and monitoring systems must be established to enable you to control the work. With a large number of labourers to supervise, effective implementation usually means that the construction operations need to be broken down into clearly defined work activities which are easy to plan, easy to instruct, and easy to supervise. To establish and then continuously monitor *realistic* labour productivity rates is usually the key to proper planning, supervision and costing of labour-based works.

In order to assess the volume of work involved in a road maintenance contract and to control completed work, a labour-based road maintenance contractor does not need to master advanced survey and setting-out techniques. However, the contractor should make sure that appropriate measuring aids, like profile boards and slope templates, are available for the required setting-out works.

### REMEMBER

○ **Supervision** of larger groups of labour requires specialist skills.
○ A key to achieving higher productivity is the introduction of an incentive scheme adapted to the task.
○ When establishing a planning and monitoring system, it must be simple enough for the supervisors to have time to fill it in, yet detailed enough to provide the necessary information.

# Part 1 – Business Questions

This section will help you to analyse how well prepared you and your company are to enter the labour-based road maintenance business. Go through the ten questions and answer all of them with yes or no. Then you can compare your answers with our checklist. In this checklist we help you to identify how you should prepare your company and yourself for this business opportunity.

|  | Yes | No |
|---|---|---|
| 1. Have you ever run a business with more than 50 manual labourers organized into several gangs? | ☐ | ☐ |
| 2. Have you ever run a business which relied mainly on casual labourers? | ☐ | ☐ |
| 3. Have you ever run a business where the workers are spread around different sites? | ☐ | ☐ |
| 4. Have you ever set up a camp-site for temporary accommodation of site workers? | ☐ | ☐ |
| 5. Have you ever had to set up and control a store for tools and equipment so as to minimize theft and wastage? | ☐ | ☐ |
| 6. Can you list the main operations required in road construction? | ☐ | ☐ |
| 7. Can you list the tools and equipment required for each operation? | ☐ | ☐ |
| 8. Do you know the range of task rates that you should set gang leaders according to local conditions? | ☐ | ☐ |
| 9. Have you ever been responsible for setting out a road project? | ☐ | ☐ |
| 10. Have you ever set up a planning and reporting system for a construction project? | ☐ | ☐ |

## COMMENTS TO BUSINESS QUESTIONS

How many yes answers did you give? Multiply the number by ten, and check your percentage score. Your score will tell you how strong and well prepared your company is. You may wish to look through the following checklist to help you understand why we think yes is the best answer to all these questions.

| 1., 2., 3. and 4. | Managing a large casual labour force is a special skill. Make sure you are able to do this effectively *before* you sign your name to a contract. |
| 5. | Individual hand tools are relatively cheap, but you will have to buy many of them and the cost of stolen or damaged tools could be high. |
| 6., 7. and 8. | You need to understand the basics of this business before you decide whether to invest in it. |
| 9. | Although road maintenance does not require much setting-out, this would help you to understand more about road works. |
| 10. | Good planning and reporting is vital to control widely-spaced construction or maintenance works. |

# Part 2 – Business Practice

### EXERCISE 1: PLANNING OF SITE CAMP ESTABLISHMENT FOR CONSTRUCTION WORK

The road to be improved has a length of 15km and connects the main road (village A) with villages B and C. It has to be constructed using labour-based methods.

There is sufficient labour available. You appreciate that you need to employ about 80 to 100 casual labourers for the construction work. Most of the labourers live in or near villages A, B and C.

The work on this road should start about two months from now.

*Tasks:*

1.) Plan where you would locate the site camp(s) (for construction work only). If you decide to have more than one camp or to shift a camp, then please describe when and how you would organize it and describe the reasons for your decision.

2.) Describe what your camp(s) would look like, e.g. size, facilities, personnel, etc.

41

## EXERCISE 2: CONSTRUCTION ACTIVITIES

You are a committee member of the local Contractors' Association. In the new National Programme for Rural Road Construction the road authority has issued a statement that it intends fully to involve small scale private contractors. The managers of the National Programme invited you to an information meeting where the basic organization and work proceedings of the labour-based programme were introduced using the ROMAR package of Handbook and Workbook. Subsequently, your committee organized a briefing meeting for all contractors interested in bidding for labour-based road works. Your colleagues have many questions and as you are the only member who participated in the information meeting everybody expects you to provide the answers. Are you capable of answering their questions?

A sample of their questions are listed below:

Question 1: What are the principal work operations for labour-based road construction excluding gravelling?

Question 2: What is usually the maximum labour force you need to employ on a labour-based site?

Question 3: I have currently three fairly experienced supervisors working for me. What qualities are most important when selecting one of them to lead the work at a labour-based site?

Question 4: My partner and I have been offered the opportunity to buy slightly-used surveying equipment – a theodolite and a levelling instrument – from another contractor. Although the offer is advantageous compared to the list price, they seem rather expensive. But if we put in bids for labour-based road maintenance contracts, wouldn't we need this equipment to carry out the required surveying work?

## NOW CHECK YOUR ANSWERS

Our suggested answers are at the end of this chapter. We suggest you check your answers against them before deciding on your action programme.

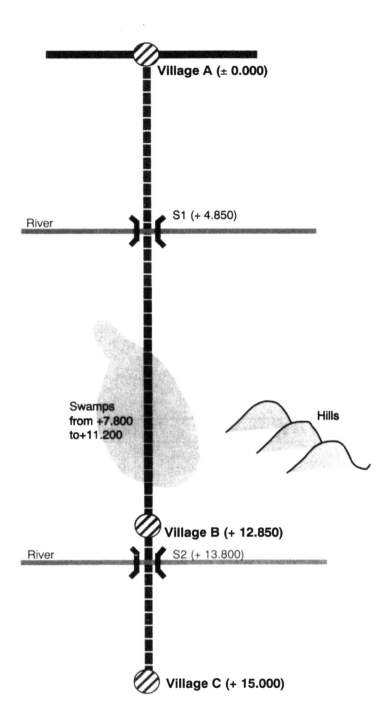

Village A (± 0.000)

River

S1 (+ 4.850)

Swamps
from +7.800
to +11.200

Hills

Village B (+ 12.850)

River

S2 (+ 13.800)

Village C (+ 15.000)

Figure 5.1

43

# Part 3 – Action Programme

## HOW TO CONSTRUCT YOUR ACTION PROGRAMME

Parts 1 and 2 should have helped you to understand your strengths and weaknesses as the owner or manager of a labour-based road maintenance enterprise. The general questions in Part 1 are a good guide to the strength of your business, and the areas where there is most room for improvement. Look back at what percentage of 'yes' answers you had; the more yes answers, the more likely it is that your business will do well.

Now look again at those questions where you answered 'no'. These may be problem or opportunity areas for your business. Choose the one which you consider most important for your business at the present time. This is the sensible way to improve your business – take the most urgent problem first; don't try to solve everything at once.

Now write the problem or opportunity into the action programme chart, as we have done with the example. Then write in *What must be done, By whom* and *By when* in order to make sure things improve.

Finally, go back to your business and carry out the action programme.

| Problem | What must be done | By whom? | By when? |
|---|---|---|---|
| I am used to setting-out for building projects, but have no equipment for setting-out road works. | I already have stringlines, tape measures and spirit levels. But I must buy 2 sets of boning rods and 10 ranging rods. My carpenter must make 4 wooden templates for side drains, 2 camber boards and a culvert bed template. | Self, Carpenter | Before first contract starts |

# Answers to Business Practice

## EXERCISE 1: PLANNING OF SITE CAMP ESTABLISHMENT FOR CONSTRUCTION WORK:

We suggest the site arrangements shown in Figure 5.2.

We intend to locate camp no. 1 at chainage +4.000. This is not too far from the river where water is available and at the same time we are just 4km from the end of the road, which is still acceptable for workers to walk.

Camp no. 2 we will locate at chainage +11.300. We would transfer the camp to this site as soon as we have reached somewhere between chainage +7.500 and +8.000 with our construction work. The camp will be located near, but outside of, village B. We assume that there is fresh water available coming from the hills and draining into the swamps.

We require accommodation for our site supervisor only. The headpersons are casually employed and will be recruited from within the vicinity of the road. In addition we require a tool store, a small site office, a pit latrine, a bath hut and a small hut for the watchman. The camp-site should be fenced with barbed wire.

For the camp we need one storeperson, three watchmen and two water carriers.

For this exercise there is not just one correct answer, as it depends very much on the local environment and the way in which you plan to carry out the work. However, what is important is that you consider all these elements and take them into consideration when planning. Always remember, labourers are human beings and not machines! If you look after their interests, they are much more likely to look after yours.

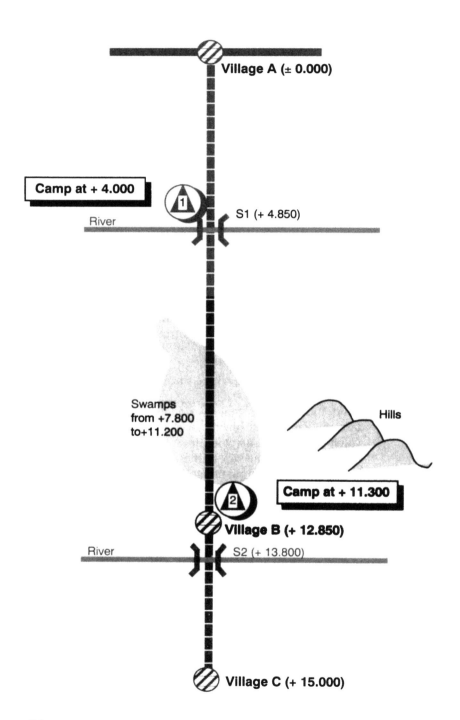

**Village A (± 0.000)**

Camp at + 4.000

S1 (+ 4.850)

River

Swamps
from +7.800
to+11.200

Hills

Camp at + 11.300

**Village B (+ 12.850)**

River

S2 (+ 13.800)

**Village C (+ 15.000)**

Figure 5.2

46

## EXERCISE 2: CONSTRUCTION ACTIVITIES

Question 1: What are the principal work operations for labour-based road construction excluding gravelling?

Our suggested answer:

The most important construction operations are:

○ supporting activities, like site store-keeping, security, water provision, setting-out
○ site clearing, like bush clearing, grass cutting, grubbing, stump and boulder removal
○ earth work (excavation), including excavation of the sub-grade and formation drainage, including side drains, mitre drains, culverts, scour checks and if necessary catch water drains
○ simple low-cost structures using locally available material.

Question 2: What is usually the maximum labour force you need to employ on a labour-based site?

Our suggested answer:

The answer to this question depends on how you decide to organize the work. It is usually best to form gangs of 10-25 labourers to undertake each clearly-defined activity. Based on the resources you have access to, the co-ordination and quality control must be undertaken by your supervisors, yourself or a combination of the two. Experiernce from various countries in running labour-based projects suggests that 80 to 100 labourers is the maximum number a trained site supervisor can control.

Question 3: I have currently three fairly experienced supervisors working for me. What qualities are most important when selecting one of them to lead the work at a labour-based site?

Our suggested answer:

The skills and qualities needed for a supervisor on any construction site (for example ability to plan, a capacity to earn the respect of the staff and of the client's supervisor, and integrity) are of course also needed on a labour-based site. A specific skill, especially useful on labour-based road works, is the ability to supervise and control large numbers of labourers.

If you intend to make a long-term commitment to enter the labour-based road sector, you should seek opportunities for your selected supervisor to attend training in the use of labour-based road maintenance and construction methods. This training can be in the form of lectures, practical site experience or ideally a combination of the two. Remember that training starts at the top, with the owner or manager, but it should then spread to the supervisors, and they should provide practical on-the-job training to the work force.

Question 4: My partner and I have been offered the opportunity to buy slightly-used surveying equipment – a theodolite and a levelling instrument – from another contractor. Although the offer is advantageous compared to the list price, they seem rather expensive. But if we put in bids for labour-based road maintenance contracts, wouldn't we need this equipment to carry out the required surveying work?

Our suggested answer:

It is not really surprising that the other contractor is willing to sell because these instruments are costly and not really necessary for labour-based works. Instead you should consider buying some 4 string line levels, 2 sets of boning rods, 10 ranging rods, 4 tape measures (30m) and 4 spirit-levels. If, in addition, you manufacture 4 wooden templates for side drains, 2 camber boards and 1 culvert bed template after you have identified the exact standards for all road elements, you will be sufficiently equipped to undertake the required surveying work.

# SECTION B
# PRACTICE

# CHAPTER 6:   ROAD MAINTENANCE

## Quick reference

All roads need maintenance to prolong their useful life, reduce vehicle operating costs and improve road transport services. As with maintenance of cars and construction plant, the owner gets the best results by undertaking regular maintenance rather than waiting for a complete breakdown. There are many ways in which roads can deteriorate, including obvious problems such as degradation of the carriageway, causing rutting or pot-holes. However, drainage problems, such as silting and erosion of the drainage system, are often more serious.

There are many causes of road deterioration, but the most important are rainfall, steep gradients, flat gradients, traffic, pavement construction and excessive growth of vegetation.

As in most conventional construction contracts, the client is usually responsible for the establishment of road inventories, the assessment of maintenance requirements, the setting of priorities and the preparation of general work-plans. The contractor's job starts when a contract is awarded, and he or she will then concentrate on the implementation of the maintenance work in accordance with the conditions of contract.

### REMEMBER

○ Because there are various ways in which a road can deteriorate, the causes must be established separately for each section of road.

○ Rutting, pot-holes and deformation of the carriageway is caused mainly by traffic.

○ Loss of gravel is caused mainly by steep gradients and rainfall.

○ Erosion of the drainage system is caused by rainfall, steep gradients and inadequate vegetation.

○ Silting of the drainage system is caused by rainfall, flat gradients, and inadequate vegetation.

○ Long-term work-plans are provided by the client, but short-term work-plans must be prepared by the contractor.

# Part 1 – Business Questions

This section will help you to analyse how well prepared you and your company are to enter the labour-based road maintenance business. Go through the ten questions and answer all of them with yes or no. Then you can compare your answers with our checklist. In this checklist we help you to identify how you should prepare your company and yourself for this business opportunity.

|  | Yes | No |
|---|---|---|
| 1. Can you state the three main purposes of road maintenance? | ☐ | ☐ |
| 2. Do you understand that you will have to spend enough time to master the technical aspects of road maintenance if you are to be successful? | ☐ | ☐ |
| 3. Have you had any techncial training in the building or construction trades? | ☐ | ☐ |
| 4. Have you or your family had any experience in road construction or maintenance? | ☐ | ☐ |
| 5. Are you willing to spend time preparing careful plans? | ☐ | ☐ |
| 6. Do you have experience of keeping records of work done? | ☐ | ☐ |
| 7. Do you have experience in working with contract documents? | ☐ | ☐ |
| 8. Are you willing to work outside normal office hours to deal with emergencies? | ☐ | ☐ |
| 9. Are you prepared to accept detailed supervision and instruction from government supervisors? | ☐ | ☐ |
| 10. Have you definitely decided to make a long-term commitment to road maintenance contracting? | ☐ | ☐ |

## COMMENTS TO BUSINESS QUESTIONS

How many yes answers did you give? Multiply the number by ten, and check your percentage score. Your score will tell you how strong and well prepared your company is. You may wish to look through the following checklist to help you understand why we think yes is the best answer to all these questions.

| | |
|---|---|
| 1. | To be a serious contractor you must understand the purpose of your business. |
| 2. and 3. | If you are not interested in the technical aspects you should choose another form of business activity. |
| 4. | Experience can help you to avoid expensive mistakes. |
| 5., 6. and 7. | If you get bored with paperwork and documents, you will not achieve long-term success as a contractor. |
| 8. | Roads are a vital public asset, and you must be prepared to put your contract commitments before personal convenience. |
| 9. | Contractors must be prepared to accept detailed supervision, and must build good relationships with supervisors even when they feel that comments are unreasonable. |
| 10. | You will not 'get rich quick' as a road maintenance contractor, although you may gradually build a worthwhile and profitable business. |

# Part 2 – Business Practice

You have been invited to tender for a routine maintenance contract for a gravel road of 25km. Before you start to prepare your offer you decide to visit the particular road to get an impression of its condition. The sketch overleaf gives you an overview of the road and the relevant main features. As you visit the road you come across a number of severe defects, and the following three exercises provide you with an opportunity to show that you have the skills to work out the causes of the problems so that you can propose sensible and cost-effective solutions to common road maintenance problems.

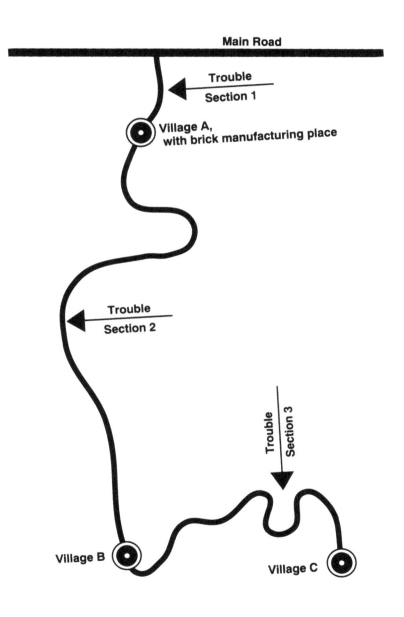

Main Road

Trouble
Section 1

Village A,
with brick manufacturing place

Trouble
Section 2

Trouble
Section 3

Village B

Village C

Figure 6.1

54

## EXERCISE 1: THE POT-HOLES AND THE KILN

The road section leading from village A to the main road has a carriageway with many pot-holes. Next to the village there is a kiln where bricks for the whole district are burnt.

Why are there so many pot-holes on this section of road?

*Your proposal:*  .......................................
.......................................
.......................................
.......................................

## EXERCISE 2: THE CASE OF THE MISSING DRAINS

This section of the road has completely silted side drains and mitre drains over a length of 3km. In some places it is not even clear where the mitre drains are, as vegetation has completely overgrown the road.

What has caused these problems?

*Your proposal:*  .......................................
.......................................
.......................................
.......................................

## EXERCISE 3: WHERE HAS ALL THE GRAVEL GONE?

In this section the side drains have deep erosion gullies and in some places the slope to the ditch has been eroded as well. On the same section there is obviously very little gravel left on the carriageway. Some deep erosion gullies run along the carriageway as well.

What has caused these failures?

*Your proposal:*  .......................................
.......................................
.......................................
.......................................

## NOW CHECK YOUR ANSWERS

Our suggested answers are at the end of this chapter. We suggest you check your answers against them before deciding on your action programme.

# Part 3 – Action Programme

## HOW TO CONSTRUCT YOUR ACTION PROGRAMME

Parts 1 and 2 should have helped you to understand your strengths and weaknesses as the owner or manager of a labour-based road maintenance enterprise. The general questions in Part 1 are a good guide to the strength of your business, and to the areas where there is most room for improvement. Look back at what percentage of 'yes' answers you had; the more yes answers, the more likely it is that your business will do well.

Now look again at those questions where you answered 'no'. These may be problem or opportunity areas for your business. Choose the one which you consider most important for your business at the present time. This is the sensible way to improve your business – take the most urgent problem first; don't try to solve everything at once.

Now write the problem or opportunity into the action programme chart, as we have done with the example. Then write in *What must be done*, *By whom* and *By when* in order to make sure things improve.

Finally, go back to your business and carry out the action programme.

| Problem | What must be done | By whom? | By when? |
|---------|-------------------|----------|----------|
| A tree has fallen and blocked the road in an area where my firm is responsible for routine maintenance. | Contact the roads supervisor for instructions if possible. Otherwise arrange labour to remove the tree and open the road, keeping a careful record of the costs involved. | Self or foreman | Immediate |

# Answers to Business Practice

## EXERCISE 1: THE POTHOLES AND THE KILN

This section carries heavy traffic coming from the kiln. The carriageway of a gravel road cannot cope with such heavy traffic, so the occurrence of potholes and ruts is not surprising. The carriageway will be further weakened during rainy periods as water will lie in the ruts and penetrate the subgrade. Regular maintenance will help, but the road authority may have to consider a more permanent solution.

## EXERCISE 2: THE CASE OF THE MISSING DRAINS

Silt is deposited where the gradient is not sufficient to allow the water to pass at a speed that allows silt to be carried away by the water flow. The area where the 3km section is located seems to be very flat. The additional reason for the silting up of the drains seems to be the lack of adequate maintenance. This is evident from the fact that most drains are overgrown by vegetation. Where drains are not regularly cleaned, growing vegetation can also hinder the free flow of the water and retain silt. Regular maintenance will help, but the road authority might save money in the long run by redesigning the drainage system so that there will be a bigger gradient to make it self-cleaning with occasional maintenance.

## EXERCISE 3: WHERE HAS ALL THE GRAVEL GONE?

The main cause of these failures is the steep gradient together with heavy rainfalls. The rate of erosion could be reduced by constructing scour checks at reasonable intervals in the side ditches.

# CHAPTER 7: ROUTINE MAINTENANCE

## Quick Reference

A routine maintenance contract requires the contractor to carry out regular work to keep a road or road network open at all times. There are four main types of routine maintenance contracts:

- single lengthperson contract
- petty contract (or labour group)
- small-scale contract for a particular road
- small-scale contract for a specified road network

The main aim of routine maintenance of earth and gravel roads is to get rainwater off the road as quickly as possible, while causing as little damage as possible. To develop work-plans you need to know the quantity of work to be carried out, the work standards to be achieved, the resources you have available, the productivity rates for each activity and the work system you have chosen.

### REMEMBER

- Priority in routine maintenance of earth and gravel roads is given to the drainage system.
- Routine maintenance activity data sheets are useful to keep records of all data that matter for planning, estimating and work monitoring.
- Reliable productivity rates are essential for operational planning and cost estimating.
- Develop a simple planning and reporting system for your contract.

# Part 1 – Business Questions

This section will help you to analyse how well prepared you and your company are to enter the labour-based road maintenance business. Go through the ten questions and answer all of them with yes or no. Then you can compare your answers with our checklist. In this checklist we help you to identify how you should prepare your company and yourself for this business opportunity.

|  | Yes | No |
|---|---|---|
| 1. In your business activities, do you prefer opportunities that offer a reasonable but regular return without high risk? | ☐ | ☐ |
| 2. Are you good at recruiting reliable workers who will perform well without detailed supervision? | ☐ | ☐ |
| 3. Are you prepared to travel long distances to supervise your work? | ☐ | ☐ |
| 4. Do you understand the basic principles of road drainage? | ☐ | ☐ |
| 5. Have you checked on minimum wage legislation for casual construction workers? | ☐ | ☐ |
| 6. Have you had experience in estimating labour inputs into construction projects? | ☐ | ☐ |
| 7. Have you checked on reasonable productivity rates in your area for common routine maintenance operations? | ☐ | ☐ |
| 8. Are you good at preparing accurate reports and accounts to describe work done? | ☐ | ☐ |
| 9. Do you know how to prepare a routine maintenance activity data sheet? | ☐ | ☐ |
| 10. Have you talked to other contractors and to supervisors from the roads authority about common problems in routine maintenance? | ☐ | ☐ |

## COMMENTS TO BUSINESS QUESTIONS

How many yes answers did you give? Multiply the number by ten, and check your percentage score. Your score will tell you how strong and well prepared your company is. You may wish to look through the following checklist to help you understand why we think yes is the best answer to all these questions.

| 1. | Road maintenance offers more reliable, but more limited, returns than new construction. |
|---|---|
| 2. | You will need to recruit a lot of workers, and mistakes will be expensive. |
| 3. | Road maintenance involves work in remote areas. |
| 4. | Well-maintained drainage is the key to a well-maintained road. |
| 5., 6. and 7. | As labour will be the main resource, you need to understand how to calculate and forecast costs. |
| 8. | You will not be paid unless you submit clear reports and accurate accounts. |
| 9. | The routine maintenance activity data sheet is the basis for your contract with the roads authority. |
| 10. | When entering a new business area, it pays to seek advice widely. You can then decide how much of it is worth taking. |

# Part 2 – Business Practice

### EXERCISE 1: ROUTINE MAINTENANCE WORK PLAN

You are issued with a contract to carry out routine maintenance of the local gravel road L 54. The road is 25km long and was constructed one year ago. The gravel course is 15cm thick (compacted) and the width of the gravelled surface is 5m.

The road is located in hilly terrain and there are very frequent rainfalls. The rainy season starts in March and ends in June.

Labour is available in plenty along the road. Based on the quantities of work mentioned in the BQ you have calculated that it will be necessary to employ casual labourers who can work every day (five days a week). Due to the high and dense rainfalls you have decided to use the lengthperson system which allows you to work continuously on all sections of the road. Each labourer is given a section of approximately 2.5km and therefore 10 lengthpersons are required in total. For the daily supervision you have employed two gangleaders, each supervising five labourers or 12.5km. The two gangleaders each have a bicycle to be able to supervise all labourers on a daily basis.

The first contract phase is for three months (13 weeks), starting from 1 April.

You are now required to prepare an overall workplan for this period. Use the form shown in Figure 7.1 to prepare your plan.

## EXERCISE 2: PREPARING THE FIRST MONTHLY ACCOUNT

At the end of the first four-week period, your achieved output is as shown in Figure 7.2. We are going to submit a monthly account to the client based on current progress. We start by filling in a schedule of 'Work carried out during account period'. The next steps, not to be done in this exercise, are to price the work according to rates agreed in the contract and present the account to the client. If you and the client agree, you should receive your monthly payment accordingly.

Your task: Fill in the table entitled 'Work carried out during account period'.

**Work carried out during account period:**

| Item No. | Description | Unit | Quantity |
|---|---|---|---|
|  |  |  |  |
|  |  |  |  |
|  |  |  |  |
|  |  |  |  |
|  |  |  |  |

## NOW CHECK YOUR ANSWERS

Our suggested answers are at the end of this chapter. We suggest you check your answers against them before deciding on your action programme.

Figure 7.1

# ROUTINE MAINTENANCE PLAN AND REPORT FOR 3 MONTHS

Road: \_\_\_\_\_  Length: \_\_\_\_ Km  Period: from \_\_\_\_ to \_\_\_\_

No. of Labourers: \_\_\_\_  Avail. Working Days: \_\_\_\_  Avail. Work-days: \_\_\_\_

| Activity | Unit | Quantity | P. Rate | WD | | Week 1 | | Week 2 | | Week 3 | | Week 4 | | Week 5 | | Week 6 | | Week 7 | | Week 8 | | Week 9 | | Week 10 | | Week 11 | | Week 12 | | Week 13 | | Total | |
|---|---|---|---|---|---|---|---|---|---|---|---|---|---|---|---|---|---|---|---|---|---|---|---|---|---|---|---|---|---|---|---|---|---|---|
| | | | | | | WD | Q | WD | Q | WD | Q | WD | Q | WD | Q | WD | Q | WD | Q | WD | Q | WD | Q | WD | Q | WD | Q | WD | Q | WD | Q | WD | Q |
| Inspection and Removal of Obstructions | m | 24 000 | | | Pl | | | | | | | | | | | | | | | | | | | | | | | | | | | | | |
| | | | | | Ach | | | | | | | | | | | | | | | | | | | | | | | | | | | | | |
| Clean Culverts and Culvert Inlets | No | 84 | | | Pl | | | | | | | | | | | | | | | | | | | | | | | | | | | | | |
| | | | | | Ach | | | | | | | | | | | | | | | | | | | | | | | | | | | | | |
| Clean Culvert Outfalls | m | 986 | | | Pl | | | | | | | | | | | | | | | | | | | | | | | | | | | | | |
| | | | | | Ach | | | | | | | | | | | | | | | | | | | | | | | | | | | | | |
| Clean Side Drains | m | 18 200 | | | Pl | | | | | | | | | | | | | | | | | | | | | | | | | | | | | |
| | | | | | Ach | | | | | | | | | | | | | | | | | | | | | | | | | | | | | |
| Repair Scour Checks | No | 62 | | | Pl | | | | | | | | | | | | | | | | | | | | | | | | | | | | | |
| | | | | | Ach | | | | | | | | | | | | | | | | | | | | | | | | | | | | | |
| Clean Mitre Drains | m | 1220 | | | Pl | | | | | | | | | | | | | | | | | | | | | | | | | | | | | |
| | | | | | Ach | | | | | | | | | | | | | | | | | | | | | | | | | | | | | |
| Fill Potholes and Ruts | m2 | 110 | | | Pl | | | | | | | | | | | | | | | | | | | | | | | | | | | | | |
| | | | | | Ach | | | | | | | | | | | | | | | | | | | | | | | | | | | | | |
| Grab Carriageway Edge | m | 5800 | | | Pl | | | | | | | | | | | | | | | | | | | | | | | | | | | | | |
| | | | | | Ach | | | | | | | | | | | | | | | | | | | | | | | | | | | | | |
| Cut Grass | m2 | 14 200 | | | Pl | | | | | | | | | | | | | | | | | | | | | | | | | | | | | |
| | | | | | Ach | | | | | | | | | | | | | | | | | | | | | | | | | | | | | |
| Clear Bush | m2 | 3800 | | | Pl | | | | | | | | | | | | | | | | | | | | | | | | | | | | | |
| | | | | | Ach | | | | | | | | | | | | | | | | | | | | | | | | | | | | | |
| Miscellaneous | - | - | | | Pl | | | | | | | | | | | | | | | | | | | | | | | | | | | | | |
| | | | | | Ach | | | | | | | | | | | | | | | | | | | | | | | | | | | | | |
| Total | | | | | Pl | | | | | | | | | | | | | | | | | | | | | | | | | | | | | |
| | | | | | Ach | | | | | | | | | | | | | | | | | | | | | | | | | | | | | |

P. Rate = Productivity Rate (Task Rate)    WD = Work-days    Q = Quantity    Pl = Planned Output    Ach = Achieved Output

62

# Part 3 – Action Programme

## HOW TO CONSTRUCT YOUR ACTION PROGRAMME

Parts 1 and 2 should have helped you to understand your strengths and weaknesses as the owner or manager of a labour-based road maintenance enterprise. The general questions in Part 1 are a good guide to the strength of your business, and to the areas where there is most room for improvement. Look back at what percentage of 'yes' answers you had; the more yes answers, the more likely it is that your business will do well.

Now look again at those questions where you answered 'no'. These may be problem or opportunity areas for your business. Choose the one which you consider most important for your business at the present time. This is the sensible way to improve your business – take the most urgent problem first; don't try to solve everything at once.

Now write the problem or opportunity into the action programme chart, as we have done with the example. Then write in *What must be done, By whom* and *By when* in order to make sure things improve.

Finally, go back to your business and carry out the action programme.

| Problem | What must be done | By whom? | By when? |
|---------|-------------------|----------|----------|
| My monthly account to the client has been rejected. | Improve the accuracy of my reporting system and preparation of accounts by re-reading IYCB Handbook 2 and by taking advice from my contractors' association. | Self | Before next account is due |

# ROUTINE MAINTENANCE PLAN AND REPORT FOR 3 MONTHS

Figure 7.2

Road: L 54  Length: 25.000 Km  Period: from 1.4. to 30.6.

No. of Labourers: ___  Avail. Working Days: ___  Avail. Work-days: ___

| Activity | Unit | Quantity | P. Rate | | W1 WD | W1 Q | W2 WD | W2 Q | W3 WD | W3 Q | W4 WD | W4 Q | Weeks 5–13 | Total WD | Total Q |
|---|---|---|---|---|---|---|---|---|---|---|---|---|---|---|---|
| Inspection and Removal of Obstructions | m | 25 000 | -- | PI | | | | | | | | | | | |
| | | | | Ach | 1 | | 1 | -- | 2 | -- | 1 | -- | | | |
| Clean Culverts and Culvert Inlets | No | 84 | 1 | PI | | | | | | | | | | | |
| | | | | Ach | 8 | 8 | 10 | 10 | 10 | 10 | 9 | 9 | | | |
| Clean Culvert Outfalls | m | 986 | 25 | PI | | | | | | | | | | | |
| | | | | Ach | 4 | 100 | 4 | 100 | 3 | 75 | 4 | 100 | | | |
| Clean Side Drains | m | 18 200 | 50 | PI | | | | | | | | | | | |
| | | | | Ach | 38 | 1900 | 32 | 1600 | 36 | 1800 | 35 | 1750 | | | |
| Repair Scour Checks | No | 62 | | PI | | | | | | | | | | | |
| | | | | Ach | | | - | | - | | - | | | | |
| Clean Mitre Drains | m | 1220 | | PI | | | | | | | | | | | |
| | | | | Ach | | | - | | - | | - | | | | |
| Fill Potholes and Ruts | m2 | 110 | | PI | | | | | | | | | | | |
| | | | | Ach | | | - | | - | | - | | | | |
| Grab Carriageway Edge | m | 5 800 | | PI | | | | | | | | | | | |
| | | | | Ach | | | - | | - | | - | | | | |
| Cut Grass | m2 | 14 200 | | PI | | | | | | | | | | | |
| | | | | Ach | | | - | | - | | - | | | | |
| Clear Bush | m2 | 3 800 | | PI | | | | | | | | | | | |
| | | | | Ach | | | - | | - | | - | | | | |
| Miscellaneous | - | - | | PI | | | | | | | | | | | |
| | | | | Ach | | | | | | | 1 | | | | |
| Total | | | | PI | | | | | | | | | | | |
| | | | | Ach | 51 | | 47 | | 51 | | 50 | | | | |

P. Rate = Productivity Rate (Task Rate)   WD = Work-days   Q = Quantity   PI = Planned Output   Ach = Achieved Output

# Answers to Business Practice

## EXERCISE 1: ROUTINE MAINTENANCE WORK-PLAN

There is, of course, not just one correct plan for this contract. So if your plan looks somehow different from ours it does not mean that yours must be wrong. However, there are a few important aspects that have to be considered when preparing the plan:

*Assumptions:*

○ The overall number of working days available has to be checked from the calendar.

○ The drainage activities are given highest priority and we therefore start by 'distributing' work-days on these activities. It is important that the culverts and drains are kept open throughout the contract as there are frequent rains at this time of year.

○ The other activities, especially grass cutting and bush clearing, are scheduled relatively late in the contract period, as they do not have a high priority. A second reason is that we would like to have the grass cut and bush cleaned shortly before the client comes to carry out the work measurements. If we cut the grass at the beginning of the contract then the grass will have probably grown tall again before the measurements are taken.

○ The gangleaders are supposed to carry out the daily inspection of the road. We have therefore not specifically planned for this activity. Also the activity 'miscellaneous' is open for any emergency activity, although we have planned for it towards the end of the contract. We believe that towards the end of the contract we might be in a better position to identify and suggest to the client additional jobs that need to be done.

We have now filled in the form according to the following steps:

*Step 1*

○ fill in the header of the form (Road, Length, Period)
○ enter the number of labourers working on the road excluding the gangleaders, as they are not producing work as such (10 labourers)
○ count the working days during the planning period and enter the number next to 'Avail. Working Days' (according to the calendar there are 65 working days during the contract period)
○ calculate the 'Avail. Work-days' by multiplying the 'Avail. Working Days' by the 'No. of Labourers' (65 working days × 10 labourers = 650 work-days)

*Step 2*

○ decide which productivity rates you want to use. We have checked the productivity standards provided in Chapter 7 of the Handbook and have taken the average productivity rate for planning purposes.
○ the first item 'Inspection' and the last item 'Miscellaneous' cannot be allocated a productivity rate, but will be carried out on a daily basis.

*Step 3*

○ now you have to divide the 'Quantity' of each item by the 'Productivity Rate' to get the 'Work-days' in the next column.
○ the two items 'Inspection' and 'Miscellaneous' you should fill in at the end as these items allow you to fill in 'gaps'.
○ add up all the work-days and check it against the total 'Avail. Work-days' in the header. If the total is more then you have in the header (650 work-days), then you have to revise your productivity rates until the total is less. If the total is less you can now calculate the difference between your total of the activity work-days and the available work-days in the header (in our example the difference is 27 work-days = 650 Avail. Work-days − 623 Work-days).
○ you can now distribute the remaining workdays between the two activities for which you have not yet allocated work-days (Inspection and Miscellaneous). As inspection has to be a continuous process throughout the contract period we allocated some 20 work-days to this activity while the remaining 7 work-days were allocated to 'Miscellaneous'.
○ again check the total of all work-days and make sure it agrees with the 'Avail. Work-days' from the header (650 work-days in our example).

Figure 7.3

**ROUTINE MAINTENANCE PLAN AND REPORT**  Road: *L 54*   Length: *25.000* Km   Period: from *1.4.* to *30.6.*
**FOR 3 MONTHS**

No. of Labourers: *10*   Avail. Working Days: *65*   Avail. Work-days: *650*

| Activity | Unit | Quantity | P. Rate | WD | PI/Ach | W1 | W2 | W3 | W4 | W5 | W6 | W7 | W8 | W9 | W10 | W11 | W12 | W13 | Total WD | Total Q |
|---|---|---|---|---|---|---|---|---|---|---|---|---|---|---|---|---|---|---|---|---|
| Inspection and Removal of Obstructions | m | 25 000 | -- | 20 | PI / Ach | *as and when required: removal of obstructions (inspection to be carried out on a daily basis by the gangleaders)* → | | | | | | | | | | | | 20 | 20 | 25000 |
| Clean Culverts and Culvert Inlets | No | 84 | 1 | 84 | PI / Ach | 10 | 10 | 10 | 10 | 5 | 5 | 5 | 5 | 5 | 5 | 5 | 5 | 4 | 84 | 84 |
| Clean Culvert Outfalls | m | 986 | 25 | 40 | PI / Ach | 4 | 4 | 4 | 4 | 4 | 4 | 4 | 4 | 2 | 2 | 2 | 2 | - | 40 | 986 |
| Clean Side Drains | m | 18 200 | 50 | 364 | PI / Ach | 36 | 36 | 36 | 36 | 36 | 36 | 36 | 36 | 36 | 36 | - | - | - | 364 | 18200 |
| Repair Scour Checks | No | 62 | 4 | 16 | PI / Ach | - | - | - | - | - | - | - | - | - | 7 | 9 | - | - | 16 | 62 |
| Clean Mitre Drains | m | 1220 | 50 | 25 | PI / Ach | - | - | - | - | 5 | 5 | 5 | 5 | 5 | - | - | - | - | 25 | 1220 |
| Fill Potholes and Ruts | m2 | 110 | 10 | 11 | PI / Ach | - | - | - | - | - | - | - | - | - | - | 11 | - | - | 37 | 110 |
| Grab Carriageway Edge | m | 5800 | 250 | 23 | PI / Ach | - | - | - | - | - | - | - | - | - | - | 17 | 6 | - | 23 | 5800 |
| Cut Grass | m2 | 14 200 | 350 | 41 | PI / Ach | - | - | - | - | - | - | - | - | - | - | - | 36 | 5 | 41 | 14200 |
| Clear Bush | m2 | 3 800 | 200 | 19 | PI / Ach | - | - | - | - | - | - | - | - | - | - | - | - | 19 | 19 | 3800 |
| Miscellaneous | - | - | - | 7 | PI / Ach | - | - | - | - | - | - | - | - | 2 | - | 2 | 1 | 2 | | |
| **Total** | | | | **650** | PI / Ach | 50 | 50 | 50 | 50 | 50 | 50 | 50 | 50 | 50 | 50 | 50 | 50 | 50 | 650 | |

WD = Work-days   Q = Quantity   PI = Planned Output   Ach = Achieved Output

P. Rate = Productivity Rate (Task Rate)

67

*Step 4*

○ now you have to distribute the total work-days for each activity over the contract period. As mentioned in the assumptions, we first determine the priorities for the activities. Secondly it is important not to carry out too many activities at the same time. Therefore we have identified the cleaning of culverts, the cleaning of culvert outfalls and the cleaning of side drains as our priority activities.

○ the second important planning parameter is the total number of work-days we can assume per week, which is 50 work-days. Therefore the total work-days allocated to activities cannot exceed 50 per week!

○ as we cannot plan for the activity 'Miscellaneous' we shall, for planning purposes, allocate the days wherever we have to fill a gap; we know that we shall have to carry out the work whenever the work needs to be done. Should the client order us to carry out a particular 'miscellaneous' job, then we shall either add more labourers or ask our present labourers to work more days (Saturdays and Sundays), or we shall have to re-plan some activities.

○ the 'Inspection' activity we shall, for planning purposes, allocate wherever we have a gap towards the end of the contract period. As already mentioned in the assumptions, we intend to carry out the inspection with our gangleaders as they move up and down the road anyway on a daily basis. At the same time we allocate in our plan the 20 workdays as labourer days, because the gangleaders are not included in our workforce. This has the advantage that we have a 'security buffer' of 20 work-days for which the client will pay us as an activity carried out. Through our arrangement with the gangleader we have now the possibility either to allocate the 20 work-days, or a part of it, to activities where more work is needed than we initially planned, or to carry out work which we had not originally expected. In the best case we may make an additional profit!

○ the allocation of work-days per week and activity is now a matter of juggling the figures around to make sure that the total work-days per week are reached (50 work-days), to make sure that not more than four or five activities are allocated per week, and to make sure that the total work-days per activity over the contract period is in accordance with our planned work-days.

*Step 5*

○ after all the work-days are allocated and the activity totals, as well as the weekly totals, agree with one another you need then to calculate the total quantity for each activity in the last column. Counter-check these totals against the quantity column at the front of the table (BQ quantity).

○ finally you have to counter-check once more all the figures, calculations and your assumptions!

## EXERCISE 2: PREPARING THE FIRST MONTHLY ACCOUNT

The first, and very important, step is to sum the activities we have carried out under each heading during the 4-week period. We then enter the results in column 'Total' on the right-hand side of the form.

The basis for our claim to the client looks like this:

### MONTHLY ACCOUNT

Project:                     Routine Maintenance to Road L 54
Contract Period:             2 April to 30 April

Monthly Account No.:         1
Account Period:              1 April to 30 April

**Work carried out during account period:**

| Item No. | Description | Unit | Quantity |
|---|---|---|---|
| 1 | Inspection and removal of obstructions | days | 5 |
| 2 | Clean culverts and culvert inlets | no. | 37 |
| 3 | Clean culvert outfalls | m | 375 |
| 4 | Clean side drains | m | 7050 |
| 11 | Miscellaneous* | days | 1 |

\* this work has been carried out in accordance with client's written instructions of 25 April.

Based on these quantities of work carried out, the next step is to price these activities according to the rates agreed in your contract and to submit your claim, including the table above, to the client.

Figure 7.4

**ROUTINE MAINTENANCE PLAN AND REPORT**  Road: **L 54**  Length: **25.000** Km  Period: from **1.4.** to **30.6.**
**FOR 3 MONTHS**

No. of Labourers:  Avail. Working Days:  Avail. Work-days:

| Activity | Unit | Quantity | P. Rate | WD | | Week 1 WD | Week 1 Q | Week 2 WD | Week 2 Q | Week 3 WD | Week 3 Q | Week 4 WD | Week 4 Q | Week 5–13 | Total WD | Total Q |
|---|---|---|---|---|---|---|---|---|---|---|---|---|---|---|---|---|
| Inspection and Removal of Obstructions | m | 25 000 | -- | | PI | | | | | | | | | | | O |
| | | | | | Ach | 1 | -- | -- | 1 | 2 | -- | -- | 1 | | 5 | - |
| Clean Culverts and Culvert Inlets | No | 84 | 1 | | PI | | | | | | | | | | | O |
| | | | | | Ach | 8 | 8 | 10 | 10 | 10 | 10 | -- | 9 | | 37 | 37 |
| Clean Culvert Outfalls | m | 986 | 25 | | PI | | | | | | | | | | | O |
| | | | | | Ach | 4 | 100 | 4 | 100 | 3 | 75 | 4 | 100 | | 15 | 375 |
| Clean Side Drains | m | 18 200 | 50 | | PI | | | | | | | | | | | O |
| | | | | | Ach | 38 | 1900 | 32 | 1600 | 36 | 1800 | 35 | 1750 | | 141 | 7050 |
| Repair Scour Checks | No | 62 | | | PI | | | | | | | | | | | |
| | | | | | Ach | - | | - | | - | | - | | | | |
| Clean Mitre Drains | m | 1220 | | | PI | | | | | | | | | | | |
| | | | | | Ach | - | | - | | - | | - | | | | |
| Fill Pot-holes and Ruts | m2 | 110 | | | PI | | | | | | | | | | | |
| | | | | | Ach | - | | - | | - | | - | | | | |
| Grab Carriageway Edge | m | 5800 | | | PI | | | | | | | | | | | |
| | | | | | Ach | - | | - | | - | | - | | | | |
| Cut Grass | m2 | 14 200 | | | PI | | | | | | | | | | | |
| | | | | | Ach | - | | - | | - | | - | | | | |
| Clear Bush | m2 | 3800 | | | PI | | | | | | | | | | | |
| | | | | | Ach | - | | - | | - | | - | | | | |
| Miscellaneous | - | - | | | PI | | | | | | | | | | | |
| | | | | | Ach | | | | | | | 1 | | | | |
| **Total** | | | | | Ach | 51 | | 47 | | 51 | | 50 | | | | |

P. Rate = Productivity Rate (Task Rate)  WD = Work-days  Q = Quantity  PI = Planned Output  Ach = Achieved Output

70

# CHAPTER 8:   REGRAVELLING

## Quick Reference

The local Roads Authority should have a long-term plan covering periodic maintenance activities, like regravelling. Depending on several factors, like weather conditions, traffic volume and design standard the Authority determines how often each road needs to be regravelled (usually every 5 to 10 years).

Gravelling is a complicated operation where you have to co-ordinate a large number of labourers, working on several different activities, with all the hauling equipment. This equipment, whether hired or owned by yourself, represents a substantial expenditure to your company. The work should therefore be planned so as to ensure maximum output from the equipment. When planning a regravelling contract, it is useful to divide the gravelling work into two main groups: the preparation activities (for example reshaping of the road, preparation of the quarry, and excavation and stockpiling of gravel), and the actual gravelling operation (consisting, for example, of loading and hauling of gravel, off-loading and spreading and compaction). Proper preparations are a prerequisite for achieving high productivity during the actual regravelling. The regravelling works should always be guided by the equipment operations, and the labour force must be allocated to match the output of the equipment.

On a temporary operation like regravelling you normally employ all the labourers needed, and probably some supervision staff, on a temporary basis from within the local community.

### REMEMBER

o An important factor, often determining whether you will achieve high output from the equipment (= profit on your contract), is the quality and timing of the preparation activities.

o You must be fully aware of all relevant local labour laws, such as minimum wages, before hiring your labourers.

o You must pay all labourers correctly and on time.

o Because gravelling very much depends on the equipment output you can achieve, you have to establish a good back-up for emergency repairs before you start the activities.

# Part 1 – Business Questions

This section will help you to analyse how well prepared you and your company are to enter the labour-based road maintenance business. Go through the ten questions and answer all of them with yes or no. Then you can compare your answers with our checklist. In this checklist we help you to identify how you should prepare your company and yourself for this business opportunity.

| | Yes | No |
|---|---|---|
| 1. Have you experience of project work in another area of construction, such as building or public works? | ☐ | ☐ |
| 2. Do you own a pick-up truck? | ☐ | ☐ |
| 3. Do you own any transport equipment, such as a tipper or a tractor and trailer? | ☐ | ☐ |
| 4. Do you own a pedestrian-controlled vibrating roller? | ☐ | ☐ |
| 5. Do you employ any supervisory staff with regravelling or general roads experience? | ☐ | ☐ |
| 6. Have you any experience with routine maintenance of roads? | ☐ | ☐ |
| 7. Do you have a basic understanding of the various types of soils? | ☐ | ☐ |
| 8. Do you know how to test a soil sample to make sure it is suitable for regravelling? | ☐ | ☐ |
| 9. Do you know how to check that the camber of the road surface is correct? | ☐ | ☐ |
| 10. Do you know how to prepare and submit interim and final accounts? | ☐ | ☐ |

## COMMENTS TO BUSINESS QUESTIONS

How many yes answers did you give? Multiply the number by ten, and check your percentage score. Your score will tell you how strong and well prepared your company is. You may wish to look through the following checklist to help you understand why we think yes is the best answer to all these questions.

| 1. | Running project work is different from routine maintenance work, because operations are always changing and the supervisor must be adaptable. |
|---|---|
| 2., 3. and 4. | Your new investment will not be so great if you already own basic equipment, and know how to use it. |
| 5. and 6. | You will be more likely to succeed if you and your key staff understand the business. |
| 7., 8. and 9. | You need basic technical knowledge if you are to control the work properly and understand instructions given by the client's representative. |
| 10. | In order to get paid promptly you need to prepare accounts in the proper format. |

# Part 2 – Business Practice

### EXERCISE 1: BILL OF QUANTITIES

You are asked to submit a tender for a regravelling contract. Before you calculate the unit prices you have decided to visit the road in order to prepare a general work-plan which would assist you in developing your prices.

The road is 8.100km long and connects the main road with the village. The gravelled carriageway is 4.50m wide and the gravel thickness has been specified in the tender document to be a compacted layer of 12cm. The quarry has been selected by the client and its location can be seen from the strip map. The road is traversing a relatively flat area. Only the quarry is located at a hill top. The gravel is of good quality and in agreement with the standards specified in the contract. For excavation by labour the gravel can be described as 'hard gravel'. The quarry is 85m long and 40m deep. There is an approximately 25cm thick overburden with some small trees, bushes and grass growing on it.

Your inspection of the road and the quarry has shown that the access road to the quarry needs some considerable repair work. There are also two steep sections of approximately 500m length each which require gravelling (about 3.5m wide) to allow the equipment to pass (assume a 13cm thick layer of gravel before compaction).

The road itself is still in good shape. The camber at some places is too flat, and drains are generally filled with silt.

You own 4 tractors with 8 trailers, each of which can carry 3m³. In addition, you have two pedestrian-vibrating rollers and a 2000 litre waterbowser with a spraying bar which can be towed by a tractor or strong pick-up.

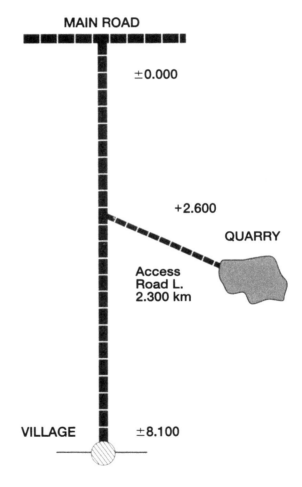

Figure 8.1

Based on the information provided, fill in the quantity column of the Bill of Quantities.

## Bill of Quantities

| Item | Description | Unit | Quantity | Rate | Amount |
|------|-------------|------|----------|------|--------|
| 1 | Establishment | lump sum | | | |
| 2 | Improvement of quarry access roads including maintenance throughout the contract period | m | | | |
| 3 | Quarry preparation; consisting of bush clearing, grass cutting, grubbing and clearing well of the quarry | m² | | | |
| 4 | Excavation of overburden including loading, hauling and stocking within 100m | m³ | | | |
| 5 | Reshaping of road, consisting of re-establishment of carriageway crossfall, reshaping of shoulders and slopes and cleaning of ditches, mitre drains and culverts | m | | | |
| 6 | Excavation of *in situ* gravel (hard) and stockpiling ready for loading (loose) | m³ | | | |
| 7 | Loading of loose gravel on to trailer or lorry | m³ | | | |
| 8 | Hauling of gravel to site; average haul distance 4.300km | m³ | | | |
| 9 | Off-loading and spreading of gravel material to the required thickness | m³ | | | |
| 10 | Watering and compacting of gravel layer by vibrating roller to the required camber at OMC ±2% in layers of no more than 120mm compacted in carriageway | m² | | | |
| | Total | | | | |
| 11 | Contingencies, | | | | |
| | Total incl. contingencies | | | | |

## EXERCISE 2: PREPARE A WORK-PLAN

Starting with the quantities calculated, and the additional information provided in Exercise 1, fill in all items in the work-plan below. This information is necessary when pricing your bid. The productivity tables presented in the Handbook should be used when calculating. They are brought together in Tables 8.1 to 8.5.

**Work-plan:**

Road length:

Carriageway:

Road condition:

Quarry access road:

Quarry:

Gravelling sections:

Total gravel required:

Total workdays required:

Total gravelling days required:

Labourers required:

Time frame:

**Table 8.1** Proposed task rates for preparation activities

| Activity | Task Rate | |
|---|---|---|
| Preparation of access road | 50–60m/work-day | |
| Clearing bush | 200–1000m²/work-day | |
| Excavating overburden and loading on to wheelbarrow | 2–4m³/work-day | |
| Hauling overburden by wheelbarrow[1] | Quantity | No. of trips/day |
| 0–40m | 10.5m³/work-day | 210 |
| 40–60m | 8.0m³/work-day | 160 |
| 60–80m | 6.5m³/work-day | 130 |
| 80–100m | 5.5m³/work-day | 110 |

[1] task rate for hauling and tipping only; excludes loading and spreading; two wheelbarrows assigned to each hauling labourer; good haul route.

**Table 8.2** Proposed task rates for road reshaping

| Activity | Task rate |
|---|---|
| Reshaping road | 20–50m/work-day |

**Table 8.3** Proposed task rates for gravel excavation

| Activity | Task rate |
|---|---|
| Excavation loose gravel | 2–3m³/work-day |
| Excavation hard gravel | 1.5–2m³/work-day |
| Excavation gravel with oversize | 1–1.2m³/work-day |
| Excavation very hard gravel with much oversize | 1m³/work-day |

**Table 8.4** Proposed task rates for loading gravel

| Activity | Task rate |
|---|---|
| Loading gravel to tractor trailer | 8–10m³/work-day |
| Loading gravel to lorry | 5–8m³/work-day |

**Table 8.5** Proposed task rates for off-loading and spreading

| Activity | Task rate |
|---|---|
| Off-loading and spreading | 12–16m³/work-day |

## NOW CHECK YOUR ANSWERS

Our suggested answers are at the end of this chapter. We suggest you check your answers against them before deciding on your action programme.

# Part 3 – Action Programme

## HOW TO CONSTRUCT YOUR ACTION PROGRAMME

Parts 1 and 2 should have helped you to understand your strengths and weaknesses as the owner or manager of a labour-based road maintenance enterprise. The general questions in Part 1 are a good guide to the strength of your business, and to the areas where there is most room for improvement. Look back at what percentage of 'yes' answers you had; the more yes answers, the more likely it is that your business will do well.

Now look again at those questions where you answered 'no'. These may be problem or opportunity areas for your business. Choose the one which you consider most important for your business at the present time. This is the sensible way to improve your business – take the most urgent problem first; don't try to solve everything at once.

Now write the problem or opportunity into the action programme chart, as we have done with the example. Then write in *What must be done, By whom* and *By when* in order to make sure things improve.

Finally, go back to your business and carry out the action programme.

| Problem | What must be done | By whom? | By when? |
|---|---|---|---|
| On my previous project, I lost money because the spreading and compacting team were waiting for material. | Improve planning to make sure that capacity for excavation, loading and transport matches capacity for spreading and compacting. | Self/Site supervisor | For next contract |

# Answers to Business Practice

## EXERCISE 1: BILL OF QUANTITIES

| Item | Description | Unit | Quantity | Rate | Amount |
|------|-------------|------|----------|------|--------|
| 1 | Establishment | lump sum | | | |
| 2 | Improvement of quarry access roads including maintenance throughout the contract period | m | 2300 | | |
| 3 | Quarry preparation; consisting of bush clearing, grass cutting, grubbing and clearing well of the quarry | m² | 3400 | | |
| 4 | Excavation of overburden including loading, hauling and stocking within 100m | m³ | 850 | | |
| 5 | Reshaping of road, consisting of re-establishment of carriageway crossfall, reshaping of shoulders and slopes and cleaning of ditches, mitre drains and culverts | m | 8100 | | |
| 6 | Excavation of *in situ* gravel (hard) and stockpiling ready for loading (loose) | m³ | 5468 | | |
| 7 | Loading of loose gravel on to trailer or lorry | m³ | 5468 | | |
| 8 | Hauling of gravel to site; average haul distance 4.300km | m³ | 5468 | | |
| 9 | Off-loading and spreading of gravel material to the required thickness | m³ | 5468 | | |
| 10 | Watering and compacting of gravel layer by vibrating roller to the required camber at OMC ±2% in layers of no more than 120mm compacted in carriageway | m² | 36 450 | | |
| | Total | | | | |
| 11 | Contingencies, | | | | |
| | Total incl. contingencies | | | | |

*Calculations*

Item 2:   the length of the quarry access road in metres = 2300m

Item 3   the area of the quarry (85m × 40m) = 3400m²

Item 4   the area of the quarry multiplied by the overburden thickness (3400m² × 0.25m) = 850m³

Item 5   the length of the road to be gravelled = 8100m

Item 6   the total volume of loose gravel (loosening factor 1.25) required for the gravel layer (8100m length × 4.50m width × 0.12m thickness × 1.25 loosening factor) = 5467.5m³

Item 7   same as item 6

Item 8   same as item 6

Item 9   same as item 6

Item 10   the gravelled surface area of the road (8100m × 4.50m) = 36 450m²

## EXERCISE 2: MY REGRAVELLING WORK-PLAN

Here is the work-plan, with all the items properly filled in. It is very useful, when pricing your bid, to have a comprehensive work-plan like this. It facilitates the calculations significantly as all the information necessary is available at a glance. It also forms the first step of planning the contract work where you arrive at overall figures on the total number of work-days needed, an overall time frame for the contract work, etc. This exercise of 'building in your mind' is most useful and provides an opportunity to save significant amounts of money (it is cheaper to make mistakes on your desk while planning than when out on the site regravelling).

*Work-plan*
Road length:   8.100km

Carriageway:   4.5m wide, gravel layer with standard thickness of 12cm (compacted, crossfall 5%).

Road condition:   generally good, some sections with flat camber, drains silted.

Quarry access road:   requires quite some repair; two sections of approx. 500m each with a width of 3.5m require gravelling. Total access road length 2.3km.

Quarry:   85m by 40m, overburden 25cm thick with some small trees, bush and grass. The gravel quality is of the required standard.

Gravelling sections:   For our plan we are dividing the road into two sections; the first section from chainage 0.000 to +2.600, and the second section from +2.600 to +8.100.

Total gravel required:   4.5m road width with a gravel thickness of 12cm compacted (+ 25% for loose material) multiplied by 8100m = 5470m³. In addition we require about 450m³ of gravel for the improvement of the steep sections of the quarry access road (3.5m width, 10cm thickness and 1000m length). The total gravel requirements are therefore about 6000m³.

Total work-days required:   (see calculations below) =
6257 work-days

Total gravelling days required:   (see calculations below) =
331 gravelling days

Labourers required:

○ Gang I; access road and quarry preparation = 70 labourers (later move to gang II)
○ Gang II; quarry work (excavation, loading) = 70 labourers
○ Gang III; road reshaping = 4 labourers
○ Gang IV; gravelling work (off-loading, spreading) Maximum labour force: = 3 to 8 labourers
○ Support Gang; for camp work, water carriers we require about 10 labourers
○ Total; when the work is in full operation we require about 95 to 100 labourers.

Time frame:

○ Site camp establishment     = 1 week
○ Quarry preparation         = 4 weeks
○ Gravelling operation       = 17 weeks
○ Total time required        = 22 weeks

*Calculation: Tractor days*
*Section 1*

○ average hauling distance = (2300 + 2600/2) = 3600m
○ amount of gravel to be hauled = 2600m × 4.50m × 0.12m × 1.25 = 1755m$^3$
○ trailer capacity = 3m$^3$
○ total trailer loads required = 1755m$^3$/3m$^3$ = 585 trips
○ trips per tractor per day (poor to good haul route) = 8 trips
○ total tractor days required for section I = 585 trips/8 trips = 74 days

*Section II*

○ average hauling distance = (2300 + 5500/2) = 5050m
○ amount of gravel to be hauled = 5500m × 4.50m × 0.12m × 1.25 = 3715m$^3$
○ trailer capacity = 3m$^3$
○ total trailer loads required = 3715m$^3$/3m$^3$ = 1240 trips
○ trips per tractor per day (poor to good haul route) = 6 trips
○ total tractor days required for section II = 1240 trips/6 trips = 207 days

Total tractor days required (section I, 74 days + section II, 207 days) = 281 days

In practice equipment can usually not be fully utilized (100%). There are almost always small breakdowns, delays in loading and off-loading, poor weather conditions, and so on that do not allow you to produce at maximum level. Therefore it is necessary to take these delays into account when calculating the time required to commplete the work. In our example we assume, based on previous experience, that we will be able to utilize the tractors at an effective rate of 85%.

With an effective utilization rate for the tractors of 85% the total becomes (281/0.85) = 331 gravelling days.

*Calculations: Work-days*
*Quarry preparation*

- access road reshaping = 2300m/40m p. wd =                    58 wd
- access road gravelling = 450m$^3$
  - excavation        450m$^3$/1.7m$^3$ p. wd =              265 wd
  - loading        450m$^3$/7m$^3$ p. wd =               65 wd
  - off-loading, spreading 450m$^3$/15m$^3$ p. wd =        30 wd
- bush clearing, grass c. = 3400m$^2$/300m$^2$ p. wd =        12 wd
- overburden removal = 850m$^3$/3m$^3$ p. wd =              284 wd
- hauling overburden average 50m, 850m$^3$/8m$^3$ p. wd = 107 wd

*Road reshaping*

- reshaping = 8100m / 30m p. wd =                            270 wd

*Gravelling*

- excavation hard gravel = 5500m$^3$/1.7m$^3$ p. wd =      3236 wd
- loading to trailers = 5500m$^3$ / 7m$^3$ p. wd × 1.18 =   928 wd
- off-loading + spreading = 5500m$^3$/15m$^3$ p. wd × 1.18 = 433 wd

Miscellaneous (support, etc.) + 10% of 5688 wd =             569 wd

Total work-days required                                    6257 wd

Due to the reduced utilization of the tractors (85%, see explanations under 'Tractor days' above) we also need to consider a lower 'utilization' of labourers carrying out activities directly related to the hauling. In practical terms you would have to allocate labourers to loading as well as off-loading and spreading, on the assumption that we would achieve the maximum possible number of tractor trips per day. However, a major breakdown affecting one of the tractors would probably make it almost impossible to achieve the maximum number of trips on which we based our calculations. If the number of trips is reduced because we have only three tractors instead of four, the labourers will not be able to work at maximum level either, i.e. we will have an under-utilization of the labour as well. In order to take this reduction of utilization into account we assume the same utilization rate for the labourers as for the tractors, 85%. This gives us a multiplier factor of 1.18 for the workerdays (100% / 85% = 1.18). This factor is employed above for the loading and the off-loading activities, while the excavation is not affected since you can normally stockpile gravel if there are not enough trailers to load due to breakdown of tractors.

## LABOURERS REQUIRED

*Quarry preparation (Gangs I + II)*
For reshaping the access road we would initially require a gang of some 10 labourers. But as we need to gravel part of the access road we will first concentrate on opening the quarry. For the first two weeks we therefore employ some 70 labourers to open the quarry and to reshape the access road. This labour force can afterwards continue with gravel excavation and stockpiling and will also be sufficient to cater for the average equipment output (32 trips a day = 8 trips per tractor with 4 tractors).

*Road reshaping (Gang III)*
The average speed that can be expected from the available equipment can be assumed from the average number of trips per day. On section I this would be 32 trips, while on section II it would be 24 trips. If we start reshaping well ahead of the actual gravelling operations we could assume an average of 28 trips per day.

With one trailer load we can gravel a road length of 4.40m (3m$^3$ / 4.50m / 0.15) with 28 trips of gravel we can therefore gravel 120m of road length per day. With a task rate of 30m / wd we therefore require a reshaping gang of 4 labourers only.

*Gravelling (Gang IV)*
The size of this gang will change in accordance with the number of trips expected to be done per day. The task rate is 15m$^3$ / wd so the minimum gang size will be 3 labourers when we are able to undertake about 12 trips or 36m$^3$ per day. The maximum gang size will be 8 labourers when we expect to do about 40 trips or 120m$^3$ per day.

## TIME REQUIRED:

○ site camp establishment with 10 labourers =          1 week
○ access road improvement and quarry opening with
                              70 labourers =   2 weeks
○ gravel excavation and stockpiling with 70 labourers = 2 weeks
○ gravelling operations with about 100 labourers
                    (see calculations below) =   17 weeks

*Gravelling calculations*
Section I: 74 trips with 4 tractors at an average utilization rate of 85% = 22 days
Section II: 207 trips with 4 tractors at an average utilization rate of 85% = 61 days
Total number of days: 22 + 61 = 83 days (in our programme we use 17 weeks = 85 days).

85

# CHAPTER 9:   PRICING AND BIDDING

## Quick Reference

Many small-scale contractors have, as you know, serious problems surviving. Experience from several countries shows that one of the most important factors, in many cases the most important, contributing to the problems is the contractor's inability to assess costs and prepare a proper bid. A typical small-scale contractor, with limited financial resources, needs to have new projects coming in at fairly regular intervals and must make a reasonable profit on every project undertaken. Putting in bids that are too high means not getting any projects while just one bid that is too low might lead to bankruptcy.

Although it is always you, the contractor, who must decide what price you think the client will accept, your decision can only be made rationally if you know your own costs for undertaking the job. To ensure that no individual cost items are forgotten, it is best to work according to a system. It will help if you list and calculate cost in an organized way. The ROMAR series follows the Improve Your Construction Business (IYCB) method that has been tested in several countries and for several subsectors. When starting to use a proper estimating method you might find it rather time-consuming (although the individual calculations often are fairly simple) but you should remember three things:

1) you will need much less time for estimating once you get used to the system;
2) without knowing your costs you cannot take sound business decisions;
3) there is no short-cut or easy way to proper estimating, all costs have to be calculated and included – 'guesstimates' very often lead to bankruptcies.

In order to prepare a bid you need information concerning both the project (bidding documents + site inspection), and your own company (company records, follow-up of previous contracts) plus your own experience. .

When pricing a routine maintenance contract the contractor normally has to assess the amount of work to be done to keep the road up to a certain standard. On a normal regravelling contract the client supplies you with a Bill of Quantities indicating the volume of work to be done.

## REMEMBER

○ If you have queries about anything in the tender or contract documents, always seek assistance from the client or your contractors' association.
○ Always make sure that all instructions for changes and extra work are given in writing.
○ Remember always to include all the indirect costs in every bid you prepare. A job that covers only the direct costs but not the indirect cost is run at a loss.

# Part 1 – Business Questions

This section will help you to analyse how well prepared you and your company are to enter the labour-based road maintenance business. Go through the ten questions and answer all of them with yes or no. Then you can compare your answers with our checklist. In this checklist we help you to identify how you should prepare your company and yourself for this business opportunity.

|  | Yes | No |
|---|---|---|
| 1. Do you always visit the site before bidding for a job? | ☐ | ☐ |
| 2. Do you have a standard checklist for site inspections? | ☐ | ☐ |
| 3. Do you always check that you will be able to recruit enough local labour at the wage rate you have used in your estimate? | ☐ | ☐ |
| 4. If you are responsible for providing the regravelling material, do you always check that there is a quarry close to the site and that you will be given a permission to excavate? | ☐ | ☐ |
| 5. If you have to supply materials do you always obtain a quotation from a dealer before bidding? | ☐ | ☐ |
| 6. If you need to hire equipment do you always check the hourly, daily or weekly rate before bidding? | ☐ | ☐ |

|        | Yes | No |
|--------|:---:|:--:|

7. Do you keep accurate cost records for your own plant, and do you use them to calculate rates when estimating? ☐ ☐

8. Do you always check the contract documents to be sure that there are no unusual provisions which could lead to increased risk? ☐ ☐

9. Before deciding to bid, do you always prepare a cash flow forecast based on a realistic forecast of when the client will pay? ☐ ☐

10. Do you always check the actual unit costs on your projects, so that your estimates gradually become more accurate? ☐ ☐

## COMMENTS TO BUSINESS QUESTIONS

How many Yes answers did you give? Multiply the number by ten, and check your percentage score. Your score will tell you how strong and well prepared your company is. You may wish to look through the following checklist to help you understand why we think Yes is the best answer to all these questions.

1. and 2.   It is far too risky to submit a bid without first checking on local conditions and trying to identify any problems you might encounter.

3.   Labour will be your major resource, so you must check that your costs are realistic.

4.   If the Roads Authority is not responsible for supplying material, you could run into difficult negotiations with land owners unless you have an agreement in advance.

5. and 6.   Dealers in remote areas often enjoy a monopoly, so you will be wise to negotiate a firm quotation of unit prices in advance.

7.   Equipment is expensive to buy, operate and maintain. If your rates are too low, it will never pay for itself.

8.   Clients are not obliged to draw your attention to unusual provisions in contract documents, so never assume that they are all the same.

9.   If you run out of cash you will lose the contract and all the savings that you have built up from previous activities. Some clients are slow to pay, so do not

assume that they will always abide by the contract conditions.

10. Never forget the estimate once you have won the contract. It gives you a guideline for planning and checking your operations. If actual productivity rates are reasonable, major differences between forecast and actual costs mean that you need to change your standard unit rates.

# Part 2 – Business Practice

### EXERCISE 1: MY OWN CHECKLIST TO MANAGE CONTRACTS

To be able to manage contracts effectively it is important to be aware of the most important procedures. Chapter 9 of the Handbook provides a very comprehensive list of procedures and activities for all kinds of contracts.

In this exercise you are asked to develop your own checklist for contract procedures and activities that you think are required for the type of contracts you expect to get.

### EXERCISE 2: MY OFFER FOR A ROUTINE MAINTENANCE CONTRACT

| | |
|---|---|
| Road name: | Piruba – Ansora |
| Road length: | 31.510km |
| Carriageway: | 5.00m wide, gravel layer with standard thickness of 15cm (compacted), crossfall 5% |
| Cross-section: | |

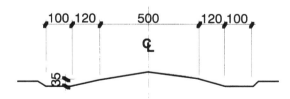

Figure 9.1

| | |
|---|---|
| Climate: | much rain in March to June and little rain in October and November (total approx. 2200mm p.a.) |
| Labour available: | average of 300 habitants per km$^2$. |

| General condition: | good condition, no major faults, drains well established but generally silted, one steep section with eroded side drain, uniform thickness of gravel layer. You are free to choose the work system (labour organization) which suits you and the local conditions best. |
| --- | --- |
| Contract period: | 6 months (1 January to 30 June). |
| Company costs: | you have two other ongoing contracts (about the same size as this contract) during the period, so this contract will keep the company one-third engaged. One of the other contracts is a rehabilitation of a school in a village located next to this road. |
| | The total company costs per year (including foremen) are estimated at 3000 NU. |
| Cost of insurance: | Contractor's full comprehensive insurance 200 NU |
| Available Resources: | 1 Road Foreman, 1 pickup (0.30 NU / km, average distance per site visit 80km), 1 motorcycle (0.15 NU / km, average daily distance 40 km), |

| Daily wage rates: | ○ labour | = 1.20 NU |
| --- | --- | --- |
| | ○ gangleader | = 2.90 NU |
| Prices for tools: | ○ hoe | = 10.00 NU |
| | ○ shovel | = 13.00 NU |
| | ○ pick axe | = 18.00 NU |
| | ○ rake | = 9.00 NU |
| | ○ grass slasher | = 7.00 NU |
| | ○ bush knife | = 9.50 NU |
| | ○ axe | = 16.00 NU |
| | ○ wheelbarrow | = 62.00 NU |
| | ○ spirit level | = 9.00 NU |
| | ○ line level | = 5.00 NU |
| | ○ ranging rode | = 14.00 NU |
| | ○ bicycle | = 450.00 NU |

Any other information required to calculate your prices can be found in the ROMAR Handbook.

*Bill of Quantities:*

| Item | Description | Unit | Quantity | Rate | Amount |
|------|-------------|------|----------|------|--------|
| 1 | Inspection and removal of obstructions and debris over the entire road length | day | 48 | | |
| 2 | Clear silt and debris from culvert, and culvert-inlet and dispose of this material safely beside side drain twice during the contract period | No. | 98 | | |
| 3 | Clear silt and debris from culvert outlet drains and dispose of safely behind drain twice during the contract period | m | 2010 | | |
| 4 | Clean the side drains to the standard cross-section removing all soil, vegetation and other debris and dispose of safely inside or outside the road reserve | m | 33 500 | | |
| 5 | Repair scour checks to standard dimensions using stones or wooden sticks | No. | 45 | | |
| 6 | Clean the mitre drains from silt and other debris and dispose of safely behind drain twice during the contract period | m | 2050 | | |
| 7 | Fill pot-holes and ruts with approved material from stockpiles and compact well | m² | 260 | | |
| 8 | Grub carriageway edge and repair erosion gullies on the shoulder | m | 12 400 | | |
| 9 | Cut grass on shoulders, ditch bottoms, side slopes and 2m from the outer side of ditch into road reserve twice during the contract period | m² | 28 000 | | |
| 10 | Clear bush and remove roots and stumps from side ditch, shoulder, slope and road reserve 2m from outer side of ditch | m² | 2500 | | |
| 11 | mergency maintenance and miscellaneous, 10% of total | | | | |
| | Total | | | | |

## EXERCISE 3: MY OFFER FOR A REGRAVELLING CONTRACT

In Chapter 8, Exercise 2, you prepared a work-plan for a re-gravelling contract. Now you are asked to prepare a complete tender for this contract. To remind you about the relevant data required to calculate the prices, here is the most important information as provided in Chapter 8.

*Situation*
The road to be regravelled is the 8.100km-long feeder road which connects the main road with the village at the end of the road. The width of the gravelled carriageway is 4.50m and the gravel thickness has been specified in the tender document to be a compacted layer of 12cm.

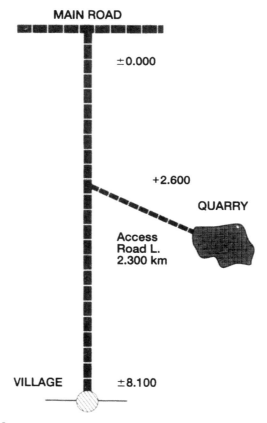

Figure 9.2

The quarry has been selected by the client and its location can be seen from the strip map. The road is traversing a relatively flat area but the quarry is located at a hill top. The gravel is of good quality and in agreement with the standards specified in the contract. For excavation by labour the gravel can be described as 'hard gravel'. The quarry is 85m long and 40m deep. The over-burden is approximately 25cm thick and there are some small trees, bushes and grass growing on it.

Your inspection of the road and the quarry has shown that the access road to the quarry needs some considerable repair work. There are also two steep sections of approximately 500m length each of which require gravelling (about 3.5m wide) to allow the equipment to pass.

The road to be gravelled is in relatively good shape although the camber at some places is too flat, and drains are generally filled with silt.

You own 4 tractors with 8 trailers, each of which can carry 3m$^3$. In addition you have two pedestrian vibrating rollers and a 2000 litre waterbowser with a spraying bar which can be towed by a tractor or strong pick-up.

*Our General Work-plan*
Here is the work-plan we developed in Chapter 8, Exercise 2, of the Workbook:

| | |
|---|---|
| Road length: | 8.100km |
| Carriageway: | 4.50m wide, gravel layer with standard thickness of 12cm (compacted, crossfall 5%). |
| Road condition: | generally good, some sections with flat camber, drains silted. |
| Quarry access road: | requires quite some repair; two sections of approx. 500m each with a width of 3.5m require gravelling. Total access road length is 2.3 km. |
| Quarry: | 85m by 40m, overburden 25cm thick with some small trees, bush and grass. The gravel quality is of the required standard. |
| Gravelling sections: | For our plan we are dividing the road into two sections; the first section from chainage 0.000 to +2.600, and the second section from +2.600 to +8.100. |
| Total gravel required: | 4.5m road width with a gravel thickness of 12cm compacted (+ 25% for loose material) multiplied by 8100m = 5470m$^3$. In addition, we require about 450m$^3$ of gravel for the improvement of the steep sections of the quarry access road (3.5m width, 10cm thickness and 1000m length). The total gravel requirements are therefore about 6000m$^3$. |

| Total work-days required: | (see calculations in Chapter 8) = 6257 work-days |
|---|---|
| Total gravelling days required: | (see calculations in Chapter 8) = 331 gravelling days |
| Labourers required: | Gang I; access road and quarry preparation = 70 labourers (later move to gang II) |
| | Gang II; quarry work (excavation, loading) = 70 labourers |
| | Gang III; road reshaping = 4 labourers |
| | Gang IV; gravelling work (off-loading, spreading) Maximum labour force: = 3 to 8 labourers |
| | Support Gang; for camp work, water carriers we require about 10 labourers |
| | Total; when the work is in full operation we require about 95 to 100 labourers. |
| Time frame: | Site camp establishment = 1 week |
| | Quarry preparation = 4 weeks |
| | Gravelling operation = 17 weeks |
| | Total time required = 22 weeks |
| Available resources: | Supervision 2 road foremen |
| | Equipment – 1 pickup (average distance per site visit – 80 km) |
| | – 4 tractors (60HP) |
| | – 8 trailers, each 3m³. |
| | – 2 pedestrian vibrating rollers |
| | – one 2000 litre waterbowser with a spraying bar |

*Cost of Labour, Equipment and Tools*

| Wage rates: | labour | 1.20 NU/day |
|---|---|---|
| | gangleader | 2.90 NU/day |
| | foreman | 3500 NU/year |
| Cost of Equipment: | pedestrian vibrating roller, per hour | 8 NU |
| | tractor with 2 trailers, per hour | 18 NU |
| | water bowser, per hour | 0.80 NU |
| | pick-up, per hour | 5 NU |

| | | |
|---|---|---|
| Company costs, per year: | Director's salary | 5000 NU |
| | Administration | 1500 NU |
| | Mechanic | 3400 NU |
| | Store keeper | 2000 NU |
| | Bookkeeper, Auditor | 300 NU |
| | Interest on bank loan | 600 NU |
| Tools: | hoe | 10.00 NU |
| | pick axe | 18.00 NU |
| | shovel | 13.00 NU |
| | mattock | 18.00 NU |
| | wheelbarrow | 62.00 NU |
| | bush knife | 9.50 NU |
| | grass slasher | 7.00 NU |
| | rake | 9.00 NU |
| | spreader | 17.00 NU |
| | earth rammer | 25.00 NU |
| | sledge-hammer | 22.00 NU |
| | crow bar | 25.00 NU |
| | spirit level | 9.00 NU |
| | sets of boning rod | 6.00 NU |
| | ranging rod | 27.00 NU |
| | ditch template | 6.00 NU |
| | line and level | 5.00 NU |
| | camber board | 10.00 NU |
| | strings | 50.00 NU |

Other information: Site office costs = 440 NU/year

Compensation to land-owner where you set up site camp = 200 NU

Contractor's full comprehensive insurance = 350 NU

We have no other contract going on at the same time.

*Bill of Quantities*

The Bill of Quantities prepared in Chapter 8, Exercise 1, looks as follows:

| Item | Description | Unit | Quantity | Rate | Amount |
|------|-------------|------|----------|------|--------|
| 1 | Establishment | lump sum | | | |
| 2 | Improvement of quarry access roads including maintenance throughout the contract period | m | 2300 | | |
| 3 | Quarry preparation; consisting of bush clearing, grass cutting, grubbing and clearing well of the quarry | m² | 3400 | | |
| 4 | Excavation of overburden including loading, hauling and stocking within 100m | m³ | 850 | | |
| 5 | Reshaping of road, consisting of re-establishment of carriageway crossfall, reshaping of shoulders and slopes and cleaning of ditches, mitre drains and culverts | m | 8100 | | |
| 6 | Excavation of *in situ* gravel (hard) and stockpiling ready for loading (loose) | m³ | 5469 | | |
| 7 | Loading of loose gravel on to trailer or lorry | m³ | 5468 | | |
| 8 | Hauling of gravel to site; average haul distance 4.3km | m³ | 5468 | | |
| 9 | Off-loading and spreading of gravel material to the required thickness | m³ | 5468 | | |
| 10 | Watering and compacting of gravel layer by vibrating roller to the required camber at OMC ±2% in layers of no more than 120mm compacted in carriageway | m² | 36 450 | | |
| | TOTAL | | | | |
| 11 | Contingencies, 10% of total | | | | |
| | FINAL TOTAL | | | | |

For any other information you require to calculate your costs, use the charts and tables available in the Handbook.

Now prepare your bid!

## NOW CHECK YOUR ANSWERS

Our suggested answers are at the end of this chapter. We suggest you check your answers against them before deciding on your action programme.

# Part 3 – Action Programme

## HOW TO CONSTRUCT YOUR ACTION PROGRAMME

Parts 1 and 2 should have helped you to understand your strengths and weaknesses as the owner or manager of a labour-based road maintenance enterprise. The general questions in Part 1 are a good guide to the strength of your business, and to the areas where there is most room for improvement. Look back at what percentage of 'yes' answers you had; the more yes answers, the more likely it is that your business will do well.

Now look again at those questions where you answered 'no'. These may be problem or opportunity areas for your business. Choose the one which you consider most important for your business at the present time. This is the sensible way to improve your business – take the most urgent problem first; don't try to solve everything at once.

Now write the problem or opportunity into the action programme chart, as we have done with the example. Then write in *What must be done*, *By whom* and *By when* in order to make sure things improve.

Finally, go back to your business and carry out the action programme.

| Problem | What must be done | By whom? | By when? |
|---------|-------------------|----------|----------|
| I have done some routine maintenance work, but am not sure how to break down work to tender for projects. | Study IYCB Handbook 1 and Workbook 1. If necessary ask for advice from quantity surveyor before preparing first bid. | Self | |

# Answers to Business Practice

### EXERCISE 1: MY OWN CHECKLIST TO MANAGE CONTRACTS

There is not a rigid list of procedures and activities available which is universally valid for the management of contracts, as the kind and nature of contracts as well the work environment, will change from case to case. Our proposal provides a list of procedures and activities that seem to us to be the most important. If your list looks similar than you are well prepared for any future contract.

*Before tendering*

○ Register the company and obtain a licence

*Tendering*

○ If anything in the tender document is not 100% clear seek outside assistance (e.g. from client directly or from contractors' association).
○ Before preparing the tender, visit the site, understand the requirements for each component and take note of the most important features.
○ Bring any perceived tender document omissions to the notice of the client and obtain clarification.
○ Decide on an overall strategy to organize and carry out the work and draw up a preliminary work programme.

*Estimating*

○ Make sure all prices are known: manpower, equipment, material.
○ Make sure all cost items are calculated: direct costs, indirect costs, profit and contingencies.
○ Check for possibilities where certain activities or resources could be combined or utilized in a more economical way.

*Financing*

○ Compare the amount of the required finances at any time with the likely income at that same time.
○ Investigate possible sources of finance to fill the expected gap and negotiate the availability of finance required if the contract is secured.

*Tender Submission*

○ Prepare tender documents based on similar work done previously, if possible, and ensure that all relevant data is available.
○ Carefully check that the tender meets the requirements and that all the arithmetic is correct.

*Mobilisation*

○ All preparation activities need to be carried out in time to allow work to start undisturbed and as planned.

*Work Management*

○ Review work programme and make any necessary changes in agreement with the client.
○ Record inputs and outputs each day, measure physical progress and compare it with the work programme.
○ Check productivities carefully and identify reasons for not achieving the set targets.
○ Prepare bill of quantities of work actually carried out. Inform the client immediately where deviations are apparent and, if necessary, put it in writing and inform the client of the resulting additional expenditures.
○ Adjust the programme and resource requirements if additional work is instructed by the client. Make sure instructions for additional work are given by the client in writing. Where activities are instructed which are not priced in the contract, agree with the client on an appropriate price and confirm it in writing.
○ Make timely arrangements for payment of labourers and staff.

*Payment Certificates*

○ Prepare payment certificates based on the achieved output, agree with the client, and ensure that the client signs the certificates.

*Training*

○ Make sure staff are properly trained and other manpower development measures are taken.
○ Instruct and coach supervisors on every possible occasion. Organize regular meetings and encourage staff to propose work improvements or changes.

*Disputes and Arbitration*

o Always inform the client early, and in full, of any disagreement.

o Always make sure instructions from the client which deviate from the contract are given in writing.

o Confirm agreements made during discussions or meetings with the client in minutes, and let the client sign the minutes.

## EXERCISE 2: MY OFFER FOR A ROUTINE MAINTENANCE CONTRACT

Before we calculate our prices we must prepare a list of our assumptions based on the BQ, available productivity tables and price lists.

## ASSUMPTIONS

*Approximate work-load*
Based on the quantities in the BQ and our productivity tables in Chapter 7 of the Handbook (Table 7.12) we estimate the work-days required to undertake the work. In this exercise we have used the average productivity rates but you should base your choice of rate on the actual conditions out on site as you find them on the site visit.

Activities:

| | | | |
|---|---|---|---|
| 1. Inspection and removal of obstructions | | = | 48 wd |
| 2. Clean culvert and inlets (98 No. x 2 / 1 No. p. wd) = | | | 196 wd |
| 3. Clean outlets (2010m x 2 / 40m p. wd) | | = | 101 wd |
| 4. Clean side drains (33 500m / 45m p. wd) | | = | 744 wd |
| 5. Repair scour checks (45 No. / 5 No. p. wd) | | = | 9 wd |
| 6. Clean mitre drains (2050m x 2 / 45m p. wd) | | = | 91 wd |
| 7. Fill pot-holes (260m² / 6m² p. wd) | | = | 44 wd |
| 8. Grub edge (12 400m / 200m p. wd) | | = | 62 wd |
| 9. Cut grass (28 000m² x 2 / 300m² p. wd) | | = | 187 wd |
| 10. Clear bush (2500m² / 250m² p. wd) | | = | 10 wd |

Total:                                                          = 1492 wd

Including miscellaneous work items we shall make arrangements to have a capacity to undertake approximately 1600 work-days.

*Assumed organization*
Due to the wet climate and the amount of work to be carried out along the entire road we chose the lengthperson system. Labour seems to be available along the road when we need it.

Based on the expected workload over 6 months of 1600 work-days we need to employ enough labour to cover approximately 270 work-days per month. Assuming that one labourer can effectively control a road section of 1.5km we require 21 labourers (31.5km / 1.5km). Each labourer would therefore have to work for 13 days per month or about 3 days per week. This seems a good system as the labourers can look after their farms and/or domestic work during the remaining weekdays.

The road foreman will be in charge of the overall supervision. He or she will be given a motorcycle enabling them to travel along the road and supervise all labourers more or less daily. We do not intend to employ gangleaders as the foreman can easily supervise 21 labourers. To ease supervision and to make the foreman more effective we will have one group of 10 labourers working on Mondays, Wednesdays and Fridays and a second group of 11 labourers working on Tuesdays, Thursdays and Saturdays. The foreman can also take care of the first item in the BQ as long as there is only inspection work to do. If any obstructions need to be removed he can always instruct the nearest labourer to undertake the job.

*Tools and equipment required*
Standard hand tool set for each labourer:

| | | |
|---|---|---|
| 1 hoe | = | 10.00 NU |
| 1 shovel | = | 13.00 NU |
| 1 rake | = | 9.00 NU |
| 1 grass slasher | = | 7.00 NU |
| 1 bush knife | = | 9.50 NU |
| Total cost per set | = | 48.50 NU |
| Total cost for 21 sets | = | 1020.00 NU |

Additional hand tools for labourers to share:

| | | |
|---|---|---|
| 10 wheelbarrows (62 NU x 10) | = | 620.00 NU |
| 5 long-handled spades (15 NU x 5) | = | 75.00 NU |
| 5 long-handled trowels (15 NU x 5) | = | 75.00 NU |
| 10 pickaxes (18 NU x 10) | = | 180.00 NU |
| 5 axes (16 NU x 5) | = | 80.00 NU |
| Total cost for additional tools | = | 1030.00 NU |

Measuring aids:

| | | |
|---|---|---|
| 2 spirit levels (9 NU x 2) | = | 18.00 NU |
| 2 sets of boning rods (6 NU x 2) | = | 12.00 NU |
| 10 ditch templates (6 NU x 10) | = | 60.00 NU |
| 2 straight edges (4 NU x 2) | = | 8.00 NU |
| 2 line levels (5 NU x 2) | = | 10.00 NU |
| strings | = | 20.00 NU |
| Total cost for measuring aids | = | 130.00 NU |
| TOTAL TOOLS AND EQUIPMENT | = | 2180.00 NU |

The tools need to be replaced after 250 working days. With 21 labourers and a foreman we can produce 5500 days of work

Daily rate for hand tools (2180 NU / 5500) = 0.40 NU

*Transport*
Pickup:
approximately 80km per site visit, each week one site visit =
27 visits = 2160 km
plus 30% for procurement of goods, visit to client, etc. = 648 km
Total: = 2808 km

Total cost for pickup (2808km x 0.30 NU/km) = 842.40 NU

This cost is shared equally with the other project in the region (school): 842.40 NU / 2 = 421.20 NU

Motorcycle:
approximately 40km per day, for 26 weeks a 6 days per week = 6240 km

Total cost for motorcycle (6240 x 0.15 NU/km) = 936.00 NU

TOTAL TRANSPORT = 1357.00 NU

*Labour wage rates*
Daily labour wage rate: = 1.20 NU

As a first step we will need to calculate the direct project costs using the IYCB standard calculation charts. To calculate the different BQ items we have to rely on our past experience and our productivity tables (task rates).

The major assumptions are stated in the following Direct Project Costs Chart:

| Direct Project Cost Chart | | | | | | | | |
|---|---|---|---|---|---|---|---|---|
| List of quantities | | | | Direct project costs (NU) | | | | |
| Item No. | Description | Unit | Quantity | Labour | Plant Tools | Material | Transp. | Total |
| 1 | Inspections and removal of obstructions and debris | day | 48 | 57.60 | 19.20 | | | 76.80 |
| 2 | Clean culverts | No. | 98 | 235.20 | 78.40 | | | 313.60 |
| 3 | Clean outlets | m | 2010 | 121.20 | 40.40 | | | 161.60 |
| 4 | Clean side drains | m | 24 350 | 649.20 | 216.40 | | | 865.60 |
| 5 | Repair scour checks | No. | 45 | 10.80 | 3.60 | | | 14.40 |
| 6 | Clean mitre drains | m | 2050 | 109.20 | 36.40 | | | 145.60 |
| 7 | Fill pot-holes | m² | 260 | 52.80 | 17.60 | | | 70.40 |
| 8 | Grub carriageway edge | m | 12 400 | 74.40 | 24.80 | | | 99.20 |
| 9 | Cut grass | m² | 28 000 | 192.00 | 64.00 | | | 256.00 |
| 10 | Clear bush | m² | 2500 | 9.60 | 3.20 | | | 12.80 |
| 11 | Emergency maintenance and miscellaneous (10%) | | | | | | | 201.60 |
| | Total direct project cost | | | | | | | 2217.60 |

*Calculations*

Item 1: Inspection and removal of debris
Assumption:  48 days indicated in the Bill of Quantities

| Cost: | labour | 48 wd x 1.20 NU = | 57.60 NU |
|---|---|---|---|
| | tools | 48 wd x 0.40 NU = | 19.20 NU |

Item 2: Clean culverts
Assumptions:  Productivity rate = 1 culvert/wd; total work-days: 98 No. x 2 / 1 No. per wd = 196 wd

| Cost: | labour | 196 wd x 1.20 NU = | 235.20 NU |
|---|---|---|---|
| | tools | 196 wd x 0.40 NU = | 78.40 NU |

Item 3: Clean culvert outlest
Assumption:   Productivity rate = 40m / wd; total work-days:
              2010m x 2 / 40m/wd = 101 wd

Cost:    labour       101 wd x 1.20 NU =              121.20 NU
         tools        101 wd x 0.40 NU =               40.40 NU

Item 4: Clean side drains
Assumption:   Productivity rate = 45m drains/wd; total work-
              days 24 350 m / 45m per wd = 541 wd

Cost:    labour       541 wd x 1.20 NU =              649.20 NU
         tools        541 wd x 0.40 NU =              216.40 NU

Item 5: Repair scour checks
Assumption:   Productivity rate = 5 No. /wd; total work-days
              45 / 5 No. per wd = 9wd

Cost:    labour       9 wd x 1.20 NU =                 10.80 NU
         tools        9 wd x 0.40 NU =                  3.60 NU

Item 6: Clean mitre drains
Assumptions:  Productivity rate = 45m/wd; total work-days:
              (2050 x 2) / 45m per wd = 91 wd

Cost:    labour       91 wd x 1.20 NU =               109.20 NU
         tools        91 wd x 0.40 NU =                36.40 NU

Item 7: Fill pot-holes
Assumptions:  Productivity rate = $6m^2$/wd; total work-days:
              $260m^2$ / $6m^2$ per wd = 44 wd

Cost:    labour       44 wd x 1.20 NU =                52.80 NU
         tools        44 wd x 0.40 NU =                17.60 NU

Item 8: Grub carriageway edge
Assumptions:  Productivity rate = 200m/wd; total work-days:
              12 400m / 200m per wd = 62 wd

Cost:    labour       62 wd x 1.20 NU =                74.00 NU
         tools        62 wd x 0.40 NU =                24.80 NU

Item 9: Cut grass
Assumption:   Productivity rate = $350m^2$/wd; total work-days:
              ($28\,000m^2$ x 2) / $350m^2$ per wd = 160 wd

Cost:    labour       860 wd x 1.20 NU =              192.00 NU
         tools        160 wd x 0.40 NU =               64.00 NU

Item 10: Clear bush

Assumption:  Productivity rate = 300m²/wd; total work-days:
2500m² / 300m² per wd = 8 wd

| Cost: | labour | 8 wd x 1.20 NU = | 9.60 NU |
| | tools | 8 wd x 0.40 NU = | 3.20 NU |

*Indirect Project Costs*

Preliminaries:

No cost for supervision is needed as we do not employ any gangleaders and the cost for the foreman is included under company costs. However, the transport costs in connection with the supervision (pick-up and motorcycle) will have to be covered here. In addition the contractor's full comprehensive insurance of 200.00 NU must be included here.

Transport for supervision (for calculation see page 103)

|  |  |
| --- | --- |
|  | 1357 NU |
| Contractor's full comprehensive insurance | 200 NU |
| Total | 1557 NU |
| Total preliminary costs | 1557.00 NU |

Risk allowance:

After judging the risks involved we arrive at a risk allowance of 4% of the total direct costs.

| Total risk allowance: 2217.60 NU x 0.04 = | 89.00 NU |
| --- | --- |

Company costs:

This contract keeps the company about one-third occupied:

| Total company costs per year | 3000 NU |
| --- | --- |
| Total company costs for 6 months | 1500 NU |
| 1/3 of total company costs for 6 months | 500 NU |
| Total company costs | 500.00 NU |

The total indirect costs are:

| Preliminaries | 1557.00 NU |
| --- | --- |
| Risk allowance | 89.00 NU |
| Company costs | 500.00 NU |
| Totals | 2146.00 NU |

The total direct costs are 2217.60 NU so the indirect costs represent approximately an additional 97% to add to the direct costs. Although this might seem a high figure, labour-based road maintenance often has very high levels of indirect costs since a lot of supervision is needed and the average daily labour rates in many countries are very low.

*Profit*
After assessing the competition for this contract and the situation of our company we have decided on a profit margin of 8% of the direct costs.

*Final cost chart*

| Item No. | Description | Direct costs NU | Indirect costs NU | Profit 8% NU | Total NU |
|---|---|---|---|---|---|
| 1 | Inspection and removal of obstructions and debris | 76.80 | 43.00 | 6.10 | 125.90 |
| 2 | Clean culverts | 313.60 | 175.60 | 25.10 | 514.30 |
| 3 | Clean culvert outlets | 161.60 | 90.50 | 13.00 | 265.10 |
| 4 | Clean side drains | 865.60 | 484.80 | 69.30 | 1419.70 |
| 5 | Repair scour checks | 14.40 | 8.10 | 1.20 | 23.70 |
| 6 | Clean mitre drains | 145.60 | 81.60 | 11.70 | 238.90 |
| 7 | Fill pot-holes | 70.40 | 39.50 | 5.70 | 115.60 |
| 8 | Grub carriageway edge | 99.20 | 55.60 | 8.00 | 162.80 |
| 9 | Cut grass | 256.00 | 143.40 | 20.50 | 419.90 |
| 10 | Clear bush | 12.80 | 7.20 | 1.00 | 21.00 |
| 11 | Miscellaneous | 201.60 | 112.90 | 16.20 | 330.70 |
| | Total | 2217.60 | 1242.20 | 177.80 | 3637.60 |

The only task remaining is to calculate and fill in the rates in the Bill of Quantities and arrive at the final offer.

*Bill of Quantities:*

| Item | Description | Unit | Quantity | Rate | Amount |
|------|-------------|------|----------|------|--------|
| 1 | Inspection and removal of obstructions and debris over the entire road length | day | 48 | 2.62 | 125.80 |
| 2 | Clear silt and debris from culvert, and culvert-inlet and dispose of this material safely beside side drain twice during the contract period | No. | 98 | 5.25 | 514.50 |
| 3 | Clear silt and debris from culvert outlet drains and dispose of safely behind drain twice during the contract period | m | 2010 | 0.13 | 261.30 |
| 4 | Clean the side drains to the standard cross section removing all soil, vegetation and other debris and dispose of safely inside or outside the road reserve | m | 24 350 | 0.059 | 1440.50 |
| 5 | Repair scour checks to standard dimensions using stones or wooden sticks | No. | 45 | 0.53 | 23.90 |
| 6 | Clean the mitre drains from silt and other debris and dispose of safely behind drain twice during the contract period | m | 2050 | 0.12 | 246.00 |
| 7 | Fill potholes and ruts with approved material from stockpiles and compact well | $m^2$ | 260 | 0.45 | 117.00 |
| 8 | Grub carriageway edge and repair erosion gullies on the shoulder | m | 12 400 | 0.013 | 161.20 |
| 9 | Cut grass on shoulders, ditch bottoms, side slopes and 2m from the outer side of ditch into road reserve twice during the contract period | $m^2$ | 28 000 | 0.015 | 420.00 |
| 10 | Clear bush and remove roots and stumps from side ditch, shoulder, slope and road reserve 2m from outer side of ditch | $m^2$ | 2500 | 0.009 | 22.50 |
| 11 | Emergency maintenance and miscellaneous, 10% of total | | | | 330.70 |
| | Total | | | | 3663.40 |

Since the rates are rounded off to a reasonable number of decimal figures (not more than 3) the figures in the 'Amount' column differ slightly from the final cost chart (for detailed explanation of the calculation steps, see Handbook chapter 9, page 238). Remember that this is not a mathematical exercise, so you decide on how to round your figures off. For example, if you feel that your rates are generally on the high side you probably round your rates downwards.

Our final offer is 3663.40 NU

## EXERCISE 3: MY OFFER FOR A REGRAVELLING CONTRACT

The first step is to calculate the direct project costs.
To be able to calculate those costs we need to know the labour rates, plant and tool costs as well as the transport costs for hauling.

The labour rate is 1.20 NU per day

**Plant and transport costs**

2 pedestrian vibrating rollers for 85 working days (17 weeks of 5 days)
(costs per hour and per roller = 8 NU)
cost per 8-hour day = 8 x 8 = 64 NU, total daily costs for 2 rollers = 128 NU

Cost for 85 days =                                              10 880 NU

4 tractors with 8 trailers for 85 working days (17 weeks of 5 days)
(costs per hour, per tractor + two trailers = 18.00 NU)
cost per 8-hour day = 8 x 18 = 144 NU
total daily costs (4 tractors/8 trailers) = 4 x 144 = 576 NU

The number of days needed for gravelling (while taking a reduced utilization rate, 85%, into account) were calculated in Exercise 2 of Chapter 8.

Cost for 85 days =                                              48 960 NU

1 water bowser, for 110 working days (22 weeks)
(cost per hour = 0.80 NU)
cost per 8-hour day = 8 x 0.80 = 6.40 NU
total daily cost = 6.40 NU

Cost for 110 days =                                               704 NU

1 pick-up for 100% of 110 working days = 110 working days
(cost per hour = 5.00 NU)
cost per 8-hour day = 40 NU, total daily cost = 40 NU

Cost for 110 days =                                                    4400 NU

**Tools**

When calculating the cost of the tools needed you start by trying
to estimate the maximum output expected during the contract.
Usually you can assume that the maximum output is deter-
mined by the haulage capacity.

How many trips per day are we planning to make, on average? When we know this we can estimate the labour requirements for excavation, loading, reshaping, off-loading and
spreading. Based on the number of labourers required we can
then determine the number of tools of each kind needed on site.

From our calculations on labour requirements we have estimated the maximum number of labourers to be about 100.
Adding some 10% for miscellaneous and support we can now
prepare a list of tools which are required for 100 labourers:

For 100 labourers we require the following tools (estimate):

| | | | |
|---|---|---|---|
| 80 hoes | @ 10.00 NU | = | 800.00 |
| 70 pick axes | @ 18.00 NU | = | 1260.00 |
| 80 shovels | @ 13.00 NU | = | 1040.00 |
| 30 mattocks | @ 18.00 NU | = | 540.00 |
| 20 wheelbarrows | @ 62.00 NU | = | 1240.00 |
| 10 bush knives | @ 9.50 NU | = | 95.00 |
| 10 grass slashers | @ 7.00 NU | = | 70.00 |
| 15 rakes | @ 9.00 NU | = | 135.00 |
| 10 spreaders | @ 17.00 NU | = | 170.00 |
| 5 earth rammers | @ 25.00 NU | = | 125.00 |
| 4 sledge-hammers | @ 22.00 NU | = | 88.00 |
| 3 crow bars | @ 25.00 NU | = | 75.00 |
| 5 spirit levels | @ 9.00 NU | = | 45.00 |
| 2 sets of boning rods | @ 6.00 NU | = | 12.00 |
| 10 ranging rods | @ 27.00 NU | = | 270.00 |
| 2 ditch templates | @ 6.00 NU | = | 12.00 |
| 2 line and levels | @ 5.00 NU | = | 10.00 |
| 2 camber boards | @ 10.00 NU | = | 20.00 |
| strings | | = | 50.00 |
| various small items | | = | 1000.00 |
| Total cost for hand tools | | = | 7057.00 NU |

We are estimating an average 1 year life span of the tools, after which we have to replace them. This gives us approximately 220 working days.

110 labourers can produce (110 labourers x 220 wd) = 24 200 work-days.

The daily rate for hand tools is therefore NU 7057 / 24 200wd = 0.30 NU

We now have all the cost elements needed and can start with our pricing exercise. Our calculation chart looks as follows:

| | | | | Direct Project Cost Chart | | | | |
|---|---|---|---|---|---|---|---|---|
| | List of quantities taken off drawings | | | | Direct project costs (NU) | | | |
| Item No. | Description | Unit | Quantity | Labour | Plant Tools | Material | Transport | Total |
| 1 | Establishment | | | 38.40 | 9.60 | | 576.00 | 624.00 |
| 2 | Improvement of access road | m | 2300 | 69.60 | 81.40 | | | 151.00 |
| 3 | Quarry preparation | m² | 3400 | 14.40 | 3.60 | | | 18.00 |
| 4 | Overburden removal | m³ | 850 | 340.80 | 85.20 | | | 426.00 |
| 5 | Reshaping of road | m | 8100 | 324.00 | 81.00 | | | 405.00 |
| 6 | Excavation of Excavation of hard gravel | m³ | 5468 | 3860.40 | 965.10 | | | 4825.50 |
| 7 | Loading | m³ | 5468 | 1106.40 | 276.60 | | | 1383.00 |
| 8 | Hauling: average distance 5.2km | m³ | 5468 | | | | 47 808.00 | 47 808.00 |
| 9 | Off-loading and spreading | m³ | 5468 | 517.20 | 129.30 | | | 646.50 |
| 10 | Watering and compaction | m² | 36 450 | 475.00 | 11 424.00 | | 2 400.00 | 14 299.00 |
| | Total direct project cost | | | 6745.60 | 13 055.80 | | 50 784.00 | 70 585.40 |
| 11 | Contingencies (10%) | | | 675.00 | 1306.00 | | 5078.00 | 7059.00 |
| | Final total direct project cost | | | 7420.60 | 14 361.80 | | 55 862.00 | 77 645.00 |

*Calculations*

Item 1: Establishment
Assumption:   Setting up and removal of 1 camp-site with 8 labourers and a tractor for 4 days

| Cost: | labour | 32 wd x 1.20 NU = | 38.40 NU |
|---|---|---|---|
| | tools | 32 wd x 0.30 NU = | 9.60 NU |
| | tractor | 4 days x 144.00 NU = | 576.00 NU |

Item 2: Improvement of access roads
Assumption:   Productivity rate = 40m/wd; total work-days: 2300m / 40m per wd = 58 wd; Water bowser needed for 2 weeks = 10 days (towed by pick-up)

| Cost: | labour | 58 wd x 1.20 NU = | 69.60 NU |
|---|---|---|---|
| | tools | 58 wd x 0.30 NU = | 17.40 NU |
| | water bowser | 10 days x 6.40 NU = | 64.00 NU |

As the pick-up will only be temporarily occupied with towing the water bowser the cost of this is not charged here. The pick-up will be included in the preliminaries unless it is fully used for a particular activity.

Item 3: Quarry preparation
Assumption:   Average production rate = 300m$^2$/wd: total work-days: 3400m$^2$ / 300m$^2$ per wd = 12 wd

| Cost: | labour | 12 wd x 1.20 NU = | 14.40 NU |
|---|---|---|---|
| | tools | 12 wd x 0.30 NU = | 3.60 NU |

Item 4: Removal of overburden
Assumption:   Productivity rate = 3m$^3$/wd; total work-days 850m$^3$ / 3m$^3$ per wd = 284 wd

| Cost: | labour | 284 wd x 1.20 NU = | 340.80 NU |
|---|---|---|---|
| | tools | 284 wd x 0.30 NU = | 85.20 NU |

Item 5: Reshaping of road
Assumption:   Productivity rate = 30m/wd; total work-days 8100 / 30m per wd = 270 wd

| Cost: | labour | 270 wd x 1.20 NU = | 324.00 NU |
|---|---|---|---|
| | tools | 270 wd x 0.30 NU = | 81.00 NU |

Item 6: Excavation of hard gravel
Assumption:   Productivity rate = 1.7m$^3$/wd; total work-days: 5468m$^3$ / 1.7m$^3$ per wd = 3217 wd

| Cost: | labour | 3217 wd x 1.20 NU = | 3860.40 NU |
|---|---|---|---|
| | tools | 3217 wd x 0.30 NU = | 965.10 NU |

Item 7: Loading

Assumption:  Productivity rate = 7m³/wd; total work-days: 5468m³ / 7m³ per wd = 781 wd x 1.18 = 922 wd

| Cost: | labour | 922 wd x 1.20 NU = | 1106.40 NU |
|-------|--------|--------------------|------------|
|       | tools  | 922 wd x 0.30 NU = | 276.60 NU  |

Due to the reduced utilization of the tractors (85%; for complete background see Chapter 8, exercise 2) we need to consider a lower 'utilization' of labourers carrying out activities directly related to the hauling. You would allocate labourers on the assumption that you will achieve the maximum number of tractor trips per day. However, since you calculate with 85% utilization of the tractors you will have a corresponding 'under-utilization' of the labour. You therefore need to assume the same 'utilization rate' for the labourers as for the tractors. In practice you multiply the work-days by a factor of 1.18 (100% / 85% = 1.18).

Item 8: Hauling average 5.2 km

Assumptions:  The work is divided into two sections, see Workbook, Chapter 8, Exercise 2. Assumed utilization rate is 85%.

A complete calculation of the resources needed to complete this activity is done in Chapter 8, Exercise 2 of this Workbook. Here is a brief summary.

The total number of tractor days needed at maximum capacity is 281. Since we have a utilization rate of 85% we arrive at 331 gravelling days (281 / 0.85 = 331). By dividing the number of gravelling days needed (331) by the number of tractors available (4) you arrive at the number of days needed for the work, 83 days.

The cost per day for 4 tractors is 576 NU (18 NU x 4 x 8 hours = 576 NU).

Cost: transport  83 days x 576 NU = 47 808.00 NU

Item 9: Off-loading and spreading

Assumption:  Productivity rate = 15m³/wd; total workdays: 5468m³ /15m³ per wd = 365 wd x 1.18 = 431 wd

| Cost: | labour | 431 wd x 1.20 NU = | 517.20 NU |
|-------|--------|--------------------|-----------|
|       | tools  | 431 wd x 0.30 NU = | 129.30 NU |

Due to the reduced utilization of the tractors (85%; for complete background see Chapter 8, Exercise 2) we need to consider a lower 'utilization' of labourers carrying out activities directly related to the hauling. You would allocate labourers on the assumption that you will achieve the maximum number of tractor trips per day. However, since you calculate with 85% utilization of the tractors you will have a corresponding under-utilization of the labour. You therefore need to assume the same 'utilization rate' for the labourers as for the tractors. In practice you multiply the workdays by a factor of 1.18 (100% / 85% = 1.18).

Item 10: Watering and compaction

Assumptions:    2 pedestrian rollers are needed for 85 days. A water bowser is needed for 85 days. A tractor or a pick-up is needed to transport water to the roadway.

The water bowser will be transported by the pick-up. If the pick-up is not available the tractors involved in the hauling could help. In order not to reduce the productivity of the hauling gang they are then paid a bonus (0.5 NU per labourer and 1 NU per tractor driver) if they achieve their hauling targets even if they have to fetch water. Assume pick-up is available for 60 of the 85 days.

2 rollers for 85 days: 8 NU x 2 x 8h x 85days =
                                                     10 880 NU
A water bowser for 85 days: 0.80 NU x 8h x
                                       85 days = 544 NU
Pick-up for 60 days: 5 NU x 8h x 60days =
                                                      2400 NU
Bonus to hauling gang when fetching water, 25 days:
30 labourers loading and spreading x 0.5 NU = 15.00 NU, 4 drivers x 1 NU = 4.00 NU, altogether 19.00 NU per day.

Cost:    labour         25 d x 19.00 NU =                475.00 NU
         equipment   10 880 NU + 544 NU =     11 424.00 NU
         transport                                          2400.00 NU

*Indirect Project Costs*

Preliminaries:

| Item | Description | Cost NU |
|------|-------------|---------|
| P1 | Contractor's site office plus furniture and all temporary buildings | 200 |
| P2 | Supervision:<br>2 Site Supervisors for 5.5 months (per year 7000 NU) = 3208 NU<br>4 gangleaders for 22 weeks = 440 days (22 x 5 x 4) @ 2.90 NU = 1276 NU | 4484 |
| P3 | Land compensation for site camps | 200 |
| P4 | Support at site camp, 8 labourers, 22 weeks at 5 days @ 1.20 NU/d | 1056 |
| P5 | Contractor's fully comprehensive insurance | 350 |
| P6 | Water bowser supplying the camp (when not used for watering the road), 0.80 NU/h x 8h x 15 days (110 – 10 – 85) = 96 NU | 96 |
| P7 | Pick-up truck, general usage (when not used on a specific activity) 5 NU/h x 8h x 50 days (110 – 60) = 2000 | 2000 |
| | Total Preliminary Costs | 8386 |

*Risk Allowance*

We are adding the risk allowance as a percentage to the total direct costs. After assessing involved risks, we decide on a risk allowance rate of 4% of the total direct costs (NU 77 644.40) which gives us 3105 NU.

Total risk allowance = 3105 NU

*Company Costs:*

To calculate the company costs we have to know the total work-load of the company at the time of this particular contract. As we have no other contract going on at the same time we assume the company costs for this contract to be 100% of the total costs.

| | |
|---|---|
| Director's salary | 5000 NU |
| Administration | 1500 NU |
| Mechanic | 3400 NU |
| Store keeper | 2000 NU |
| Bookkeeper, Auditor | 300 NU |
| Interest on bank loan | 600 NU |
| Total for one year | 12 800 NU |
| Monthly average costs (12 800 / 12) | 1067 NU |
| For this contract (1067 x 5.5 months) | 5869 NU |

The total indirect costs are:

| | |
|---|---|
| Preliminaries | 8386.00 NU |
| Risk allowance | 3105.00 NU |
| Company costs | 5869.00 NU |
| Totals | 17 360.00 NU |

The total direct costs are 77 645 NU so the indirect costs represent approximately an additional 22.5% to add to the direct costs.

*Profit*
After assessing the competition for this contract and the situation of our company we have decided on a profit margin of 6% of the direct costs.

*Total cost chart*

| Item No. | Description | Direct costs NU | Indirect costs 22.5% NU | Profit 6% | Total NU |
|---|---|---|---|---|---|
| 1 | Establishment | 624.00 | 140.40 | 37.40 | 801.80 |
| 2 | Improve access road | 150.40 | 33.90 | 9.00 | 193.30 |
| 3 | Quarry preparation | 18.00 | 4.10 | 1.10 | 23.20 |
| 4 | Overburden removal | 426.00 | 95.90 | 25.60 | 547.50 |
| 5 | Reshape road | 405.00 | 91.10 | 24.30 | 520.40 |
| 6 | Excavate hard gravel | 4825.30 | 1085.70 | 289.50 | 6200.50 |
| 7 | Loading | 1383.00 | 311.20 | 83.00 | 1777.20 |
| 8 | Hauling 3.5km | 47 808.00 | 10 756.80 | 2868.40 | 61 433.20 |
| 9 | Off-loading and spreading | 646.50 | 145.50 | 38.80 | 830.80 |
| 10 | Watering and compaction | 14 299.00 | 3217.80 | 857.90 | 18 374.70 |
| | Total | 70 584.60 | 15 882.40 | 4235.00 | 90 702.60 |
| 11 | Contingencies 10% of total | 7059.00 | 1588.00 | 424.00 | 9070.00 |
| | Total Amount, incl. contingencies | 77 643.60 | 17 470.40 | 4659.00 | 99 772.60 |

As the very last step you now have to calculate the unit rates and fill out the Bill of Quantities. We adjust the total amounts to the unit rates selected.

*Bill of Quantities*

| Item | Description | Unit | Quantity | Rate | Amount |
|------|-------------|------|----------|------|--------|
| 1 | Establishment | lump sum | | | 801.80 |
| 2 | Improvement of quarry access roads including maintenance throughout the contract period | m | 2300 | 0.084 | 193.20 |
| 3 | Quarry preparation; consisting of bush clearing, grass cutting, grubbing and clearing well of the quarry | m$^2$ | 3400 | 0.007 | 23.80 |
| 4 | Excavation of overburden including loading, hauling and stocking within 100m | m$^3$ | 850 | 0.65 | 552.50 |
| 5 | Reshaping of road, consisting of re-establishment of carriageway crossfall, reshaping of shoulders and slopes and cleaning of ditches, mitre drains and culverts | m | 8100 | 0.064 | 518.40 |
| 6 | Excavation of *in situ* gravel (hard) and stockpiling ready for loading (loose) | m$^3$ | 5469 | 1.13 | 6180.00 |
| 7 | Loading of loose gravel on to trailer or lorry | m$^3$ | 5468 | 0.33 | 1804.50 |
| 8 | Hauling of gravel to site; average haul distance 5.2km | m$^3$ | 5468 | 11.24 | 61 460.30 |
| 9 | Off-loading and spreading of gravel material to the required thickness | m$^3$ | 5468 | 0.15 | 820.20 |
| 10 | Watering and compacting of gravel layer by vibrating roller to the required camber at OMC ±2% in layers of no more than 120mm compacted in carriageway | m$^2$ | 36 450 | 0.51 | 18 589.50 |
| | Total | | | | 90 944.20 |
| 11 | Contingencies, 10% of total | | | | 9094.00 |
| | Final total | | | | 100 038.20 |

The price for the contract is thus 100 038.20 NU.

# CHAPTER 10:   MANAGING PEOPLE

## Quick Reference

*Management* can be subdivided into management of tasks and management of people. You have to be good at both of these if you are to make a success of road maintenance and regravelling. With large work-forces dispersed over wide areas, you will have to depend upon your supervisors to control the works and take decisions as soon as problems arise. A supervisor's authority comes from three sources: conferred authority, personal qualities and knowledge. You should try to recruit supervisors with the right personal qualities and with appropriate technical knowledge, but you should also take care to give them real authority and back them up when they make decisions on your behalf.

*Commnication* is the sharing of information or knowledge between two or more individuals. It is essential in managing large groups of people. Your work-force are more likely to work productivity if they feel you are telling the truth about the state of your business, and if you occasionally ask them for advice rather always than issuing orders. You may often find that they have ideas that will enable the work to be done more quickly and to a standard which pleases the client. As a manager, you should make the best use of all the skills available within the business.

*Financial rewards* are not the only incentives that matter in a private company, but an incentive bonus scheme is often effective in improving productivity, provided the extra profit is shared fairly between you and your employees.

### REMEMBER

○ The best managers are also leaders. It is sometimes said that 'a manager does things right, and a leader does the right thing.'
○ As a leader, you should always be looking at how things are done in your firm and asking yourself 'is there a better way?'
○ In order to be able to delegate responsibilities and authority a leader needs to define the objectives of the work, know the team, and be a good communicator.

# Part I – Business Questions

This section will help you to analyse how well prepared you and your company are to enter the labour-based road maintenance business. Go through the ten questions and answer all of them with yes or no. Then you can compare your answers with our checklist. In this checklist we help you to identify how you should prepare your company and yourself for this business opportunity.

|  | Yes | No |
|---|---|---|
| 1. Do you prefer a business where you have to manage people rather than manage equipment? | ☐ | ☐ |
| 2. Do your workers generally stay with you as long as you have work to offer them? | ☐ | ☐ |
| 3. Do you explain tasks clearly to your gang leaders and then leave them to manage their gangs in their own way? | ☐ | ☐ |
| 4. Do you understand that it is unfair to expect your employees to accept responsibility unless you allow them the proper authority to take decisions? | ☐ | ☐ |
| 5. Do you constantly try to improve your behaviour so as to set a good example to your workers? | ☐ | ☐ |
| 6. Do you have a reputation for keeping your promises? | ☐ | ☐ |
| 7. Do you always explain tasks patiently to your workers, and ask them questions at the end to make sure that they understand them clearly? | ☐ | ☐ |
| 8. Do you take great care in preparing written instructions to make sure that they cannot be misunderstood? | ☐ | ☐ |
| 9. Do you feel that you now have enough knowledge of the technology of labour-based road maintenance to match your conferred authority and personal qualities? | ☐ | ☐ |
| 10. Do you have a consistent policy on providing incentives for your workers? | ☐ | ☐ |

## COMMENTS TO BUSINESS QUESTIONS

How many yes answers did you give? Multiply the number by ten, and check your percentage score. Your score will tell you how strong and well prepared your company is. You may wish to look through the following checklist to help you understand why we think yes is the best answer to all these questions.

| | |
|---|---|
| 1. | If you feel more comfortable managing money and equipment, you may find managing a large labour-force too much of a strain. |
| 2. | If people want to work for you, your management style is probably good. |
| 3. and 4. | Everyone needs enough *authority* to carry out the tasks for which you have made them *responsible*. |
| 5. and 6. | Leadership by example is the most affective way of billing a good and productive team. |
| 7. | Good managers explain things clearly, and then check that the other person has really understood what has to be done. |
| 8. | Good managers can write clear instructions. |
| 9. | The ROMAR Handbook and Workbook have been written to improve your knowledge. Improving your conferred authority and personal qualities is up to you! |
| 10. | If you set up a productivity scheme and then raise the targets as soon as your workers start to earn a bonus, you will not be trusted and the incentive effect will be lost. |

# Part 2 – Business Practice

### EXERCISE 1: WHAT WENT WRONG?

Six months ago you were awarded a contract for routine maintenance of 120km of road network. You have established a gang system along the roads which is supervised by two full-time maintenance foremen, one of whom (Supervisor A) is an experienced contractor's foreman who worked in your building business while the other (Supervisor B) has recently retired from his job as a labour-based supervisor with the local roads authority. To make things easy, you split the overall contract in two, and each of the two supervisors looks after about 60km of the contract.

Now the contract is in trouble. You have had a letter from the Roads Authority complaining about the standard of work in that area of road network looked after by Supervisor A, and your accounts show that you are losing money on the other part of the contract because your monthly expenditure has been greater than the amount for which you have billed the client.

This is the first road maintenance contract you have ever been awarded, and you thought it would be easy to handle in view of your previous successes as a building contractor. In fact you thought it was necessary to make a site visit only every two months, and you were convinced that you had chosen well in selecting the two sueprvisors. Supervisor A has always done a good job for you in the past, and Supervisor B seems to know everything about the technical aspects of labour-based road-works and was highly recommended to you by the Regional Engineer.

What went wrong, and how are you going to rescue the contract, both in financial terms and in relations with your client?

## EXERCISE 2: A MANAGEMENT TWENTY QUESTIONS

In this exercise you can check your understanding of the principles of ROMAR management by filling in the missing words in this 'management twenty questions'. To get the best from this exercise, you should close your Handbook and try to work out the answers from memory and from the general meaning in the rest of the sentence. In case you have problems, the box at the end of the exercise contains all the answers in alphabetical order, so this should help you fill in any gaps.

1. For a ROMAR contractor, the task of management can be divided into two parts:
   ○ management of ——————;
   ○ management of ——————.

2. It is sometimes said that 'a manager does —————— —— —————— and a leader does —————— —————— ——————'.

3. In other words, the manager carries out —————— but the leader also thinks carefully about the —————— and tries to find the best way of reaching the ——————.

4. This means that, as a leader, you should always be looking at how things are done in your firm and asking yourself '—— —————— —————— —————— —————— —— ——————?'

5. As there are so many decisions to be made, the firm will only operate efficiently if there is a system to ensure that

each decision is made at the right level. This system will be based no the idea of —————.

6. Whenever you sign a contract with a client to undertake a project, or an annual maintenance contract, you will have the ————— for delivering work to the specific standard. Since you cannot undertake every operation yourself, you now have to ————— responsibility for some subsidiary tasks to other people.

7. This means that you will give them the power to make decisions on your behalf, and they can only do this if you also delegate to them the ————— that they need so that the people to whom they give instructions understand that you will back them up.

8. When you feel one of your subordinates has made a wrong decision, you should discuss their reasons for making the decision —————, in a ————— way, and explain why you would have acted differently.

9. In such a case your role is that of a ————— ————— ——— helping your staff to understand your priorities for the firm and the way in which it conducts its business.

10. A supervisor's authority comes from three sources:
    ○ —————;
    ○ —————; and
    ○ —————.

11. In order to be effective a manager at any level must:
    ○ understand the ————— of the work and project;
    ○ know the —————; and
    ○ be able to ————— clearly and concisely.

12. The objectives for a contract manager are to complete the work within stated targets for:
    ○ —————;
    ○ —————; and
    ○ —————.

13. No two individuals are the same – you have to ————— —— ————— ————— —————.

14. Communication can be either ————— or ————.

15. Communication in the form of information should always ————— ————— ————— ————— – ————— ————— —————.

16. It is often said that 'people who never make a mistake, never make anything.' The important thing is to ————— ———— —————.

17. There are a number of other incentives that can motivate staff to work effectively and well. The most important of these is ————— —————.

18. An incentive is offered to get someone to do undertake some ——————— and ——————— task.
19. For casual labourers there are three principal incentive payment systems:
    ○ ——————— ;
    ○ ——————— ;
    ○ ——————— .
20. The ability to rise above the hour-to-hour and day-to-day pressures of your business, to take a broader and more objective view of what is happening to your business and what will happen to affect its future performance is known as taking a ——————— ———————.

| | |
|---|---|
| authority | measurable |
| communicate | objective |
| conferred authority | objectives |
| cost | people |
| daily paid work | personal qualities |
| delegate | piece work system |
| delegation | privately |
| flow in two directions – down and up | quality |
| formal | quiet and calm |
| helicopter view | responsibility |
| informal | specific |
| instructions | task |
| is there a better way? | tasks |
| job satisfaction | task work system |
| knowledge and experience | team |
| know your work team | the right thing |
| learn from your mistakes | things right |
| management teacher | time |

## NOW CHECK YOUR ANSWERS

Our suggested answers are at the end of this chapter. We suggest you check your answers against them before deciding on your action programme.

# Part 3 – Action Programme

## HOW TO CONSTRUCT YOUR ACTION PROGRAMME

Parts 1 and 2 should have helped you to understand your strengths and weaknesses as the owner or manager of a labour-based road maintenance enterprise. The general questions in Part 1 are a good guide to the strength of your business, and to the areas where there is most room for improvement. Look back at what percentage of 'yes' answers you had; the more yes answers, the more likely it is that your business will do well.

Now look again at those questions where you answered 'no'. These may be problem or opportunity areas for your business. Choose the one which you consider most important for your business at the present time. This is the sensible way to improve your business – take the most urgent problem first; don't try to solve everything at once.

Now write the problem or opportunity into the action programme chart, as we have done with the example. Then write in *What must be done, By whom* and *By when* in order to make sure things improve.

Finally, go back to your business and carry out the action programme.

| Problem | What must be done | By whom? | By when? |
|---------|-------------------|----------|----------|
| Sometimes I order hired plant and sometimes I leave it to the site supervisor. Now the regravelling gang cannot work because there is no roller. | Prepare a simple checklist on the duties of the owner/manager and the duties of the site supervisor, to define responsibilities clearly. | Self | This week |

# Answers to Business Practice

## EXERCISE 1: WHAT WENT WRONG?

The Introduction to the Handbook stated that construction work is divided into technical aspects and managerial aspects, and noted that both of these skills are required if you are to run contracts efficiently and properly. In this case you have appointed two foremen:

Supervisor A: who is experienced in the commercial management of building contracts, but who does not have the technical knowledge necessary to run a labour-based road contract; and Supervisor B: who has many years of experience in supervising labour-based roadworks, but who is new to the private sector and lacks the cost-consciousness that is necessary to achieve profitable results.

By splitting the contract in two and putting the two supervisors in competition with each other, you have exposed their weaknesses without benefitting from their strengths. Supervisor A probably sees Supervisor B as a cast-off from the public sector who cannot cope with the stresses of running a contract, while Supervisor B probably sees Supervisor A as a brash youngster who is only interested in money and cannot be bothered to do a proper job.

So who is to blame? We feel that neither Supervisor A nor Supervisor B is really to blame, because they are doing their best but are limited by gaps in their training and experience. We believe that you as the contractor are to blame, because you have failed to take advantage of the skills that your staff possess and have also failed to help them to expand those skills to enable them to do a better job.

If you are to learn from the mistakes that have been made, it is best to make a list. This is what we suggest:

1. It should have been obvious that both of the supervisors had weaknesses as well as strengths. A competent contractor would have been aware of this at the beginning, and worked out a plan to help each of them improve.
2. Three site visits in six months are certainly not sufficient. With inexperienced staff looking after a contract in an area which is new for your firm, the two supervisors should have been visited on a weekly basis. This would have enabled you to inspect and instruct regularly and on time, to discuss problems when they happened and to get a first hand picture of the ongoing work.

3. The two supervisors could have learned a lot from each other. By putting them into direct competition, you brought about a situation in which each saw the other as a rival rather than a colleague.
4. Regular joint meetings with the supervisors would have emphasized your wish that they should work together, and you could have helped them to respect and support each other.
5. You should write to your client, explaining that you regret the problems that have arisen on the contract and have reviewed the way in which the contract is managed in order to make the best use of the varied experience of your supervisory staff.
6. In addition to encouraging the two supervisors to learn from each other and providing your own direct on-the-job coaching, you should consider asking advice from your contractors' association on formal training opportunities which will enlarge the knowledge of your staff in areas where they are weak.

## EXERCISE 2: A MANAGEMENT TWENTY QUESTIONS

These are our suggested missing words to complete the 20 statements:

1. tasks, people
2. things right, the right thing
3. instructions, task, objective
4. is there a better way?
5. delegation
6. responsibility, delegate
7. authority
8. privately, quiet and calm
9. management teacher
10. conferred authority, personal qualities, knowledge and experience
11. objectives, team, communicate
12. quality, time, cost
13. know your work team
14. formal, informal
15. flow in two directions – down and up
16. learn from mistakes
17. job satisfaction
18. specific, measurable
19. daily paid work, piece work system, task work system
20. helicopter view

www.ingramcontent.com/pod-product-compliance
Lightning Source LLC
Jackson TN
JSHW011351130125
77033JS00015B/558